# Advances in Time Series Forecasting

## Editor

**Cagdas Hakan Aladag**
*Department of Statistics*
*Hacettepe University*
*Turkey*

## Co-Editor

**Erol Eğrioğlu**
*Department of Statistics*
*Ondokuz Mayis University*
*Turkey*

# CONTENTS

**PART III: HYBRID METHODS**

# FOREWORD

Generally, advanced intelligent techniques are needed to model and solve the problems encountered in various fields and also to reach desired answers. Many institutions have used various soft computing methods to solve the problems they faced, to increase their productivity and to make strategic decisions. Hence, these methods have received more attention in recent years. In turn, practitioners and academics from various fields have been working on these approaches.

Time series forecasting is one of the most challenging contemporary tasks that are being faced in different areas. In general, different types of time series have been tried for the forecasting purpose. Unfortunately, conventional time series approaches for forecasting can be insufficient in modeling real life time series. Therefore, advanced methods such as artificial neural networks and fuzzy time series have been utilized in many applications. In this eBook, advanced forecasting approaches are described, and further explained how these approaches can be used to forecast real life time series. In particular, some new forecasting approaches are firstly introduced in this eBook. In addition, this eBook provides the background for describing new methods and improving existing advanced forecasting approaches. Dr. Cagdas Hakan Aladag and Assoc. Prof. Dr. Erol Eğrioğlu, the editors of this eBook, have made meaningful contributions to the literature regarding time series forecasting in the recent past. I believe, this eBook will be useful for both practitioners and researchers who are interested in receiving comprehensive views and insights from the variety of issues covered in this eBook.

*I. Burhan Turksen*
TOBB Economy and Technology University
Turkey

# PREFACE

Time series analysis has got attention of many researches from different fields, such as business administration, economics, public finances, engineering, statistics, econometrics, mathematics and actuarial sciences. When many organizations are planning their future, they have to forecast the future. Time series analysis has been employed by many organizations, such as hospitals, universities, companies or government organizations in order to forecast how could be the future. Therefore, many time series forecasting methods have been proposed and improved in the literature. Firstly, linear models such as Box-Jenkins methods were used in many areas of time series forecasting. Furthermore, to overcome the restriction of the linear models and to account for certain nonlinear patterns observed in real problems, some nonlinear models have been proposed in the literature. However, since these nonlinear models were developed for specific nonlinear patterns, they are not capable of modeling other types of nonlinearity in time series. In recent years, to overcome these issues, efficient soft computing techniques such as artificial neural networks, fuzzy time series and some hybrid models have been used to forecast any kind of real life time series. Both theoretical and empirical findings in the literature show that these approaches give better forecasts than those obtained from conventional forecasting methods. In addition, conventional models require some assumptions such as linearity and normal distribution cannot be utilized efficiently for some real time series such as temperature and share prices of stockholders, since this kind of series contain some uncertainty in itself. However, when soft computing methods such as neural networks and fuzzy time series are used to forecast time series, there is no need to satisfy any assumption and the time series uncertainty can be forecasted efficiently.

This eBook contains recent applications and descriptions of these effective soft computing methods. The readers can learn how these methods work and how these approaches can be used to forecast real life time series. In addition, the hybrid forecasting model approach, which is based on combining different soft computing methods to get better forecasts, is explained and at the same time, the reader can find the applications of hybrid forecasting models. The reader of this eBook can also create a new hybrid forecasting model. Although the soft computing forecasting models have many advantages, at the same time there are still some problems with their usage. These problems are pointed out in this ebook. After researchers see those problems, they make some contributions to these forecasting methods by filling some gaps to obtain better forecast results. Furthermore, some new forecasting models are introduced in the eBook.

*Cagdas Hakan Aladag*
Department of Statistics
Hacettepe University
Turkey

# List of Contributors

**Aladag, C.H.**, Ph.D., Lecturer,
*Department of Statistics, Hacettepe University, Turkey*

**Aladag, S.,** Ph.D., Candidate, Assistant Social Security Expert,
*General Director of Service Provision, Republic of Turkey Social Security Institution, Ankara, Turkey*

**Bagdadioglu, N.**, Associate Professor,
*Department of Public Finance, Hacettepe University, Ankara, Turkey*

**Basaran, A.A.**, Ph.D., Research Assistant,
*Department of Public Finance, Hacettepe University, Ankara, Turkey*

**Degirmenci, I.**, Ph.D., Candidate, Research Assistant,
*Department of Statistics, Hacettepe University, Ankara, Turkey*

**Eğrioğlu, E.**, Associate Professor,
*Department of Statistics, Ondokuz Mayis University, Turkey*

**Gunay, S.**, Professor,
*Department of Statistics, Hacettepe University, Ankara, Turkey*

**Kadilar, C.**, Professor,
*Department of Statistics, Hacettepe University, Ankara, Turkey*

**Koc, S.,** M.Sc.,
*Department of Statistics, Ondokuz Mayis University, Samsun, Turkey*

**Uslu, V.R.**, Associate Professor,
*Department of Statistics, Ondokuz Mayis University, Samsun, Turkey*

**Yolcu, U.**, Ph.D., Assistant Professor,
*Department of Statistics, Giresun University, Giresun, Turkey*

2

# CHAPTER 1

## Advanced Time Series Forecasting Methods

## Cagdas Hakan Aladag[1,*] and Erol Eğrioğlu[2]

[1]*Hacettepe University, Faculty of Science, Department of Statistics, 06800, Ankara, Turkey and* [2]*Ondokuz Mayis University, Faculty of Arts and Science, Department of Statistics, 55139, Samsun, Turkey*

**Abstract:** The researchers from various fields have been studying on time series forecasting for approximately one century in order to get better forecasts for the future. To achieve high forecast accuracy level, various time series forecasting approaches have been improved in the literature. During 1980s, some crucial developments happened and time series researches changed. More sophisticated algorithms could be improved since properties of computers were enhanced. Therefore, new time series forecasting approaches such as artificial neural networks and fuzzy time series could be proposed. In the applications, these approaches have proved its success in forecasting real life time series. In addition, hybrid forecasting methods which combine these new approaches have also been improved to obtain more accurate forecasts. In recent years, these advanced time series forecasting methods have been used to forecast real life time series and satisfactory results have also been obtained.

**Keywords:** Artificial neural networks, Fuzzy time series, Forecasting, Hybrid methods.

## 1. INTRODUCTION

It is needed to accurately forecast the future in order to make right decisions. One way to forecast the future is using time series whose observations depends on the time. Therefore, time series forecasting is an important issue in various implementation areas such as finance, management, health, tourism, energy, pollution, manufacturing and so on. And, researchers from different disciplines such as statistics, mathematics, economics, business administration and econometrics have been studying time series forecasting for approximately one century.

The major aim in the time series analysis is to forecast the future values, which have not been observed yet, of the time series accurately. Therefore, many forecasting methods have been suggested to increase forecasting accuracy in the literature. The proposed forecasting approaches range from extrapolation to fuzzy time series techniques.

The history of the time series forecasting was briefly given by Pino *et al.* [1] as follows:

"Before the early 1920s, forecasts were calculated by simply extrapolating time series. What might be dubbed as modern forecasting began in 1927, when Yule presented auto-regressive techniques to forecast the annual number of sun spots [2]. His model calculated forecasts as a weighted sum of previous data. If good performance was to be achieved from this linear system, an external factor called noise had to be catered for, as this noise affects the linear system. This linear system with noise was widely used for the next 50 years, when research culminated in the ARIMA methodology proposed by Box and Jenkins [3].

From this point onwards, strongly theory-based studies focused on non-stationary and/or non-linear series: bilinear, bi-spectral or threshold models are examples of this to name but a few [4-8].

During the 1980s, two crucial developments took place that changed time series research. On the one hand, ever increasing capacity and enhanced features of personal computers meant that much longer time series could be handled and more sophisticated algorithms could be used. This went hand in hand with a second aspect the development of machine learning techniques, such as artificial neural networks".

*Address correspondence to Cagdas Hakan Aladag:* Hacettepe University, Faculty of Science, Department of Statistics, 06800, Ankara, Turkey; E-mail: chaladag@gmail.com

Artificial neural networks have been successfully used in many applications [9]. Although artificial neural networks produce very accurate forecasts for most of the time series, the real time series data such as temperature and share prices of stockholders contain some uncertainty in itself and more proper forecasting methods are needed for such time series [10]. Therefore, after artificial neural networks started to be used for time series forecasting, a new approach which is fuzzy time series method was firstly introduced by Song and Chissom [11, 12]. Then, fuzzy time series forecasting models have drawn a great amount of attention in recent years and various models have been proposed in the literature [13].

Both theoretical and empirical findings in the literature have showed that combining different methods can be an effective and efficient way to improve forecasts [14]. Therefore, in the literature, there have been various hybrid forecasting approaches that combine both the conventional and the advanced methods [15].

Advanced time series forecasting approaches such as artificial neural networks, fuzzy time series or hybrid methods have been used widely in recent years since these approaches produces more accurate forecasts than those obtained from other conventional methods and does not require satisfying any assumptions such as linearity, normal distribution and a specific observation number.

In this chapter, the information about the artificial neural networks and the fuzzy time series is presented in Section 2 and 3. The idea underlying the hybrid forecasting approaches is given in Section 4. Finally, the last section concludes the chapter.

## 2. ARTIFICIAL NEURAL NETWORKS

Artificial neural networks are mathematical models which are inspired from the biological neural networks. Artificial neural networks compose of tree main layers which are the input, the hidden and the output layers. And, these layers include a certain number of neurons. From a mathematical point of view, artificial neural networks method can be considered universal functional approximator [16]. Moreover, artificial neural networks are non linear by nature [17], which means that they can not only correctly estimate non-linear functions, but also extract non linear elements from the data [1].

Artificial neural networks approach is a method which has been successfully used in many areas for different purposes [18]. One of these areas is time series forecasting [19]. Since artificial neural networks can model both the linear and the nonlinear structure of time series, they have attracted more and more attention from both academic researchers and industrial practitioners in recent years [20]. Artificial neural networks have been widely used to model time series in various fields of applications [21] and used as a good alternative method for both linear and non-linear time series forecasting. Zhang *et al.* [9] presented a review of the current status in applications of neural networks for forecasting.

Although artificial neural networks have been successfully used in many implementations, there still some problems with this approach [22]. Artificial neural networks can be considered as an approach composes of three main elements such as architecture structure, learning algorithm and activation function. Determining the elements of the artificial neural networks issue that affect the forecasting performance of artificial neural networks must be considered carefully [23]. Brief information about artificial neural networks was presented by Eğrioğlu *et al.* [23] as follows:

One critical decision is to determine the appropriate architecture, that is, the number of layers, number of nodes in each layers and the number of arcs which interconnect with the nodes [24]. Feed forward artificial neural network has been used in many studies for forecasting. Therefore, our focus is on the feed forward networks. Determining architecture depend on the basic problem. Since, in the literature, there are not general rules for determining the best architecture, many architecture must be examined for the correct results. Fig. **1** depicts the broad feed forward artificial neural networks architecture that has single hidden layer and single output. Other important architectures include direct connections from input nodes to output nodes. Fig. **2** depicts these architectures.

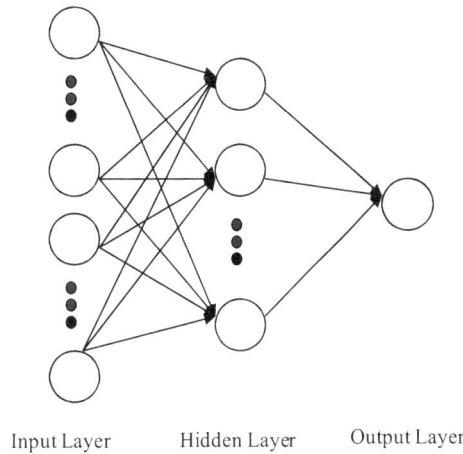

Input Layer　　　Hidden Layer　　Output Layer

**Figure 1:** A broad feed forward architecture with one output neuron.

Learning of artificial neural networks for a specific task is equivalent to finding the values all of the weights such that the desired output is generated to the corresponding input. Various training algorithms have been used for the determining optimal weights values. The most popularly used training method is the back propagation algorithm [25]. In the back propagation algorithm, learning of the artificial neural network consists of adjusting all weights such as the error measure between the desired output and actual output [26].

Another element of artificial neural networks is the activation function. It determines the relationship between inputs and outputs of a node and a network. In general, the activation function introduces a degree of the nonlinearity that is valuable for the most artificial neural networks applications. The well known activation functions are logistic, hyperbolic tangent, sine (or cosine) and the linear functions. Among them, logistic transfer function is the most popular one [9].

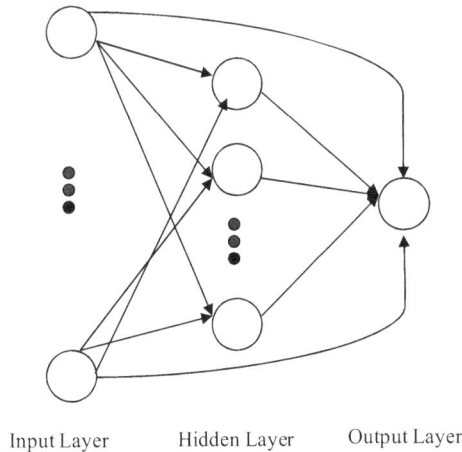

Input Layer　　　Hidden Layer　　Output Layer

**Figure 2:** A direct connected feed forward architecture with one output neuron.

## 3. FUZZY TIME SERIES

Zadeh [27] firstly introduced fuzzy set theory. Based on this paper, fuzzy set theory has found many application areas in science. Fuzzy time series approach based on fuzzy set theory was introduced as an alternative method for conventional time series models. Recently, fuzzy time series has got much attention. Song and Chissom [11, 12] first introduced fuzzy time series. These studies have been inspired by knowledge presented in the papers [28, 29]. Song and Chissom proposed a method based on matrix operations in order to forecast fuzzy time series [11, 12, 30]. Chen [31] introduced a new method, which is computationally easier than the method of Song and Chissom, by employing the fuzzy logical relationship table. Huarng and Yu [32] proposed another method in which the fuzzy logical relationships are determined by artificial neural networks. The other important studies related to fuzzy time series are conducted by Sullivan and Woodal [33], Hwang, Chen and Lee [34], Chen and Hwang [35], Huarng and Yu [36], and Yu [37, 38]. All of these studies are based on first order fuzzy time series models [10].

The first fuzzy time series forecasting model proposed by Song and Chissom [11, 12] is a one factor first order model. On the other hand, Chen [39] and Aladag *et al.* [40] presented approaches based on a one factor high order fuzzy time series forecasting model. It is likely reasonable to think that a multivariate fuzzy time series forecasting model will provide better results in terms of prediction accuracy since the real time series data can be affected by many factors [41]. In the literature, Yu and Huarng [42], and Lee *et al.* [43] used two factor fuzzy time series model to forecast such time series. Then, Jilani and Burney [44] and Eğrioğlu *et al.* [41] analyzed $k$ factor ($k \geq 2$) and $n^{\text{th}}$ order ($n \geq 1$) fuzzy time series forecasting models.

The fundamental definitions of fuzzy time series were given by Eğrioğlu *et al.* [41] as follows:

Let $U$ be the universe of discourse, where $U = \{u_1, u_2, ..., u_b\}$. A fuzzy set $A_i$ of $U$ is defined as $A_i = f_{A_i}(u_1)/u_1 + f_{A_i}(u_2)/u_2 + \cdots + f_{A_i}(u_b)/u_b$, where $f_{A_i}$ is the membership function of the fuzzy set $A_i$; $f_{A_i} : U \to [0,1]$. $u_a$ is a generic element of fuzzy set $A_i$; $f_{A_i}(u_a)$ is the degree of belongingness of $u_a$ to $A_i$; $f_{A_i}(u_a) \in [0,1]$ and $1 \leq a \leq b$.

**Definition 1.** Fuzzy time series. Let $Y(t)(t = ..., 0, 1, 2, ...)$, a subset of real numbers be the universe of discourse by which fuzzy sets $f_j(t)$ are defined. If $F(t)$ is a collection of $f_1(t), f_2(t), ...$ then $F(t)$ is called a fuzzy time series defined on $Y(t)$.

**Definition 2.** Fuzzy time series relationships assume that $F(t)$ is caused only by $F(t-1)$, then the relationship can be expressed as: $F(t) = F(t-1) * R(t, t-1)$, which is the fuzzy relationship between $F(t)$ and $F(t-1)$, where $*$ represents as an operator. To sum up, let $F(t-1) = A_i$ and $F(t) = A_j$. The fuzzy logical relationship between $F(t)$ and $F(t-1)$ can be denoted as $A_i \to A_j$ where $A_i$ refers to the left-hand side and $A_j$ refers to the right-hand side of the fuzzy logical relationship. Furthermore, these fuzzy logical relationships can be grouped to establish different fuzzy relationships.

**Definition 3.** Let $F(t)$ be a fuzzy time series. If $F(t)$ is caused by $F(t-1), F(t-2)$, ..., and $F(t-n)$, then this fuzzy logical relationship is represented by

$$F(t-n), ..., F(t-2), F(t-1) \to F(t) \tag{1}$$

and it is called the nth order fuzzy time series forecasting model.

**Definition 4.** Let F and G be two fuzzy time series. Suppose that $F(t-1) = A_i$, $G(t-1) = B_k$ and $F(t) = A_j$. A bivariate fuzzy logical relationship is defined as $A_i, B_k \to A_j$, where $A_i, B_k$ are referred to as the left hand side and $A_j$ as the right hand side of the bivariate fuzzy logical relationship.

Therefore, first order bivariate fuzzy time series forecasting model is as follows:

$$F(t-1), G(t-1) \to F(t) \tag{2}$$

**Definition 5.** Let F and G be two fuzzy time series. If $F(t)$ is caused by $(F(t-1), G(t-1)), (F(t-2), G(t-2)), ..., (F(t-n), G(t-n))$ then this fuzzy logical relationship is represented by

$$(F(t-1), G(t-1)), (F(t-2), G(t-2)), ..., (F(t-n), G(t-n)) \to F(t) \tag{3}$$

and it is called the two-factors $n$th order fuzzy time series forecasting model, where $F(t)$ and $G(t)$ are called the main factor fuzzy time series and the second factor fuzzy time series, respectively $(t = ....0, 1, 2, ....)$ .

**Definition 6.** Let $F$ and $G_1, G_2 ... G_{k-1}$ be $k$ fuzzy time series. If $F(t)$ is caused by $(F(t-1), G_1(t-1)G_2(t-1)...G_{k-1}(t-1)),...,(F(t-n), G_1(t-n)G_2(t-n)...G_{k-1}(t-n))$ then this fuzzy logical relationship is represented by

$$(F(t-1), G_1(t-1)G_2(t-1)...G_{k-1}(t-1)),...,(F(t-n), G_1(t-n)G_2(t-n)...G_{k-1}(t-n)) \rightarrow F(t) \qquad (4)$$

and it is called the $k$-factors $n$th order fuzzy time series forecasting model, where $F(t)$ and $G_i(t)$ are called the main factor fuzzy time series and the secondary factors fuzzy time series, respectively $(t = ....0, 1, 2, ...., \ i = 1, 2, ...k-1)$ .

## 4. HYBRID FORECASTING APPROACHES

Many studies available in the literature have pointed out that it is possible to obtain more accurate forecasts by using hybrid approaches combine different methods. Therefore, various hybrid forecasting approaches have been proposed in the literature in order to get better forecasts [15]. Different forecasting methods proposed in the literature can have different drawbacks or constraints. By combining different methods, it is possible to remove the relevant drawbacks or constraints. For example, by combining ARIMA models which is linear and artificial neural networks which can model nonlinear part of time series, accurate forecasts for time series include both linear and nonlinear structures can be obtained. In time series forecasting literature, some such hybrid forecasting approaches were proposed by Pai and Lin [45], Tseng *et al.* [46], Zhang [47] and Aladag *et al.* [14]. Pai and Lin [46] proposed a hybrid ARIMA and support vector machines model. Tseng *et al.* [46] combined seasonal time series ARIMA model and feed forward neural networks. Zhang [47] proposed a hybrid ARIMA and feed forward neural networks model. Aladag *et al.* [14] suggested a forecasting hybrid model composing of linear and nonlinear components as follows:

$$y_t = L_t + N_t, \qquad (5)$$

where $y_t$ denotes original time series, $L_t$ denotes the linear component and $N_t$ denotes the nonlinear component. Linear component is estimated by ARIMA model and residuals obtained from the ARIMA model

$$e_t = y_t - \hat{L}_t, \qquad (6)$$

are estimated by Elman's Recurrent Neural Networks (ERNN). Here $\hat{L}_t$ is the forecasting value for time $t$ of the time series $y_t$ by ARIMA. With $n$ input nodes, the ANN model for the residuals can be written as

$$e_t = f(e_{t-1}, e_{t-2}, \ldots, e_{t-n}) + \varepsilon_t, \qquad (7)$$

where $f$ is a nonlinear function determined by ERNN and $\varepsilon_t$ is the random error. The estimation of $e_t$ by (7) will yield the forecasting of nonlinear component of time series, $N_t$ . By this way, forecasting values of the time series are obtained by using the formula given in (4).

$$\hat{y}_t = \hat{L}_t + \hat{N}_t . \qquad (8)$$

The method proposed by Aladag *et al.* [14] is presented in below.

**Step 1.** Box-Jenkins models are used to analyze the linear part of the problem. That is, $\hat{L}_t$ is obtained by using Box-Jenkins method.

***Step 2.*** ERNN model is developed to fit the residuals from the Box-Jenkins models. That is, $\hat{N}_t$ is obtained by using ERNN.

***Step 3.*** Using (8), forecasts of the hybrid method are obtained by adding the estimates of linear and nonlinear components of the time series, found in Step 1 and Step 2, respectively.

Aladag *et al.* [14] made a comparison between their method and the other methods available in the literature and showed that the best forecasts are obtained when the method proposed by Aladag *et al.* [14] is used.

An example hybrid approach in which traditional and advanced forecasting methods are combined given above. Also, various hybrid forecasting approaches in which different advanced techniques such as heuristics algorithms, artificial neural networks, fuzzy time series and forecast combination are used exists in the time series forecasting literature. Some hybrid forecasting methods proposed by Eğrioğlu *et al.* [10], Aladag *et al.* [13], Aladag *et al.* [15], and Aladag [22] can be given as examples for such methods.

## 5. CONCLUSION

The time series forecasting is a crucial issue in many applications. And, researchers from different disciplines are working to get more accurate forecasts. There are various forecasting methods in the time series forecasting literature. These approaches can be divided into two subclasses which are conventional and advanced forecasting methods. Also, some hybrid approaches combine different methods exist in the literature and hybrid approaches can be considered as advanced forecasting approaches. The advanced time series forecasting methods have been used in many applications and have proved its success in the literature. These forecasting methods such as artificial neural networks and fuzzy time series have drawn a great amount of attention in recent years and various models have been proposed in the literature since these forecasting methods provide three crucial advantages in real life applications. These advantages can be given as follows:

- Using the advanced forecasting methods is easier than using the conventional methods.

- The advanced forecasting methods do not include assumption constraints such as linearity, normal distribution and a specific observation number.

- The advanced forecasting methods generally produce more accurate forecasts than those obtained from the conventional methods.

Therefore, using the advanced time series forecasting methods for forecasting real life time series will be wiser in order to obtain accurate forecasts.

## REFERENCES

[1]    R. Pino, J. Parreno, A. Gomez, and P. Priore, "Forecasting next-day price of electricity in the Spanish energy market using artificial neural networks," *Eng Appl Artif Intel*, vol. 21, no. 1, pp. 53-62, 2008.
[2]    G. Yule, "On a method of investigating periodicity in disturbed series with special reference to Wolfer's sunspot numbers," *Phil Trans R Soc London*, vol. 226, pp. 267–298, 1927.
[3]    G.E.P. Box, and G.M. Jenkins, *Time series analysis: Forecasting and control.* San Francisco: Holden-Day, 1970.
[4]    H. Tong, *Threshold models in non-linear time series analysis.* Springer: New York, 1983.
[5]    H. Tong, *Non-linear time series: a dynamic system approach.* Oxford University Press: Oxford, 1990.
[6]    M.B. Priestley, *Non-linear and non-stationary time series analysis.* Academic Press: London, 1988.
[7]    R.S. Tsay, "Non-linear time series analysis: diagnostics and modelling," *Stat Sinica*, vol. 1, pp. 432-451, 1991.
[8]    T.S. Rao, Analysis of Nonlinear Time Series (and Chaos) by Bispectral Methods. In: Casdagli, M., Eubank, S. (Eds.), *Nonlinear Modeling and Forecasting*, Santa Fe Institute Studies in the Sciences of Complexity Proceedings, 1992, vol. XII, pp. 199–226.

[9]    G. Zhang, and B.E. Patuwo, and Y.M. Hu, "Forecasting with artificial neural networks: the state of the art," *Int J Forecasting*, vol. 14, pp. 35-62, 1998.

[10]   E. Eğrioğlu, C.H. Aladag, V.R. Uslu, M.A. Basaran, and U. Yolcu, "A new hybrid approach based on SARIMA and partial high order bivariate fuzzy time series forecasting model," *Expert Syst Appl*, vol. 36, no. 4, pp. 7424-7434, 2009.

[11]   Q. Song, and B.S. Chissom, "Fuzzy time series and its models," *Fuzzy Set Syst.*, vol. 54, pp. 269-277, 1993.

[12]   Q. Song, and B.S. Chissom, "Forecasting enrollments with fuzzy time series-Part I," *Fuzzy Set Syst.*, vol. 54, pp. 1-10, 1993.

[13]   C.H. Aladag, U. Yolcu, and E. Eğrioğlu, "A high order fuzzy time series forecasting model based on adaptive expectation and artificial neural Networks," *Math Comput Simulat*, vol. 81, pp. 875–882, 2010.

[14]   C.H. Aladag, E. Eğrioğlu, and C. Kadilar, "Forecasting nonlinear time series with a hybrid methodology, *Appl Math Lett*, vol. 22, pp. 1467-1470, 2009.

[15]   C.H. Aladag, E. Eğrioğlu, and U. Yolcu, "Forecast combination by using artificial neural networks," *Neural Process Lett*, vol. 32, no. 3, pp. 269–276, 2010.

[16]   K. Hornik, M. Stinchcombe, and H. White, "Multilayer feed forward networks are universal approximators," *Neural Networks*, vol. 2, pp. 359–366, 1989.

[17]   D.E. Rumelhart, and J. McClelland, *Parallel Distributed Processing*, MIT Press: Cambridge, 1986.

[18]   C.H. Aladag, E. Eğrioğlu, S. Gunay, and M.A. Basaran, "Improving weighted information criterion by using optimization," *J Comput Appl Math*, vol. 233, pp. 2683-2687, 2010.

[19]   C.H. Aladag, E. Eğrioğlu, and C. Kadilar, "Modeling brain wave data by using artificial neural networks," *Hacet J Math Stat*, vol. 39, no. 1, pp. 81–88, 2010.

[20]   M. Qi, and G. Zhang, "An investigation of model selection criteria for neural network time series forecasting," *Eur J Oper Res*, vol. 132, pp. 666-680, 2001.

[21]   S. Buhamra, N. Smaoui, and M. Gabr "The Box-Jenkins Analysis and Neural Networks: Prediction and Time Series Modeling," *Appl Math Model*, vol. 27, pp. 805-815, 2003.

[22]   C.H. Aladag, "A new architecture selection method based on tabu search for artificial neural networks, *Expert Syst Appl*, vol. 38, pp. 3287-3293, 2011.

[23]   E. Eğrioğlu, C.H. Aladag, and S. Gunay, "A new model selection strategy in artificial neural network" *Appl Math Comput*, vol. 195, pp. 591-597, 2008.

[24]   J.M. Zurada, *Introduction of artificial neural systems*, St. Paul: West Publishing, 1992.

[25]   K.A. Smith, *Neural networks in business: techniques and applications*, Imprint Info Hershey: Idea Group, 2002.

[26]   A. Cichocki, and R. Unbehauen, R., *Neural networks for optimization and signal processing*, John Willey & Sons: New York, 1993.

[27]   L.A. Zadeh, "Fuzzy Sets," *Inform. and Control*, vol. 8, pp. 338-353, 1965.

[28]   L.A. Zadeh, "Outline of a new approach to the analysis of complex systems and decisions processes," *IEEE Trans. Systems Man Cybernet.*, vol. 3, pp. 28-44, 1973.

[29]   L.A. Zadeh, "The concept of a linguistic variable and its application to approximate reasoning, Parts 1-3," *Inform. Sci.*, vol. 8, pp. 199-249; pp. 301-357; vol. 9, 43-80, pp. 1975.

[30]   Q. Song, and B.S. Chissom, "Forecasting enrollments with fuzzy time series-Part II," *Fuzzy Set Syst*, vol. 62, pp. 1-8, 1994.

[31]   S.M. Chen, "Forecasting enrollments based on fuzzy time-series," *Fuzzy Set Syst*, vol. 81, pp. 311-319, 1996.

[32]   K. Huarng, and H.K. Yu, "The application of neural networks to forecast fuzzy time series," *Physica A*, vol. 363, pp. 481-491, 2006.

[33]   J. Sullivan, and W.H. Woodall, "A comparison of fuzzy forecasting and Markov modeling," *Fuzzy Set Syst*, vol. 64, pp. 279-293, 1994.

[34]   J.R. Hwang, S.M. Chen, and C.H. Lee, "Handling forecasting problems using fuzzy time series," *Fuzzy Set Syst*, vol. 100, pp. 217-228, 1998.

[35]   S.M. Chen, and J.R. Hwang, "Temperature prediction using fuzzy time series," *IEEE Trans Syst Man, Cybern B, Cybern*, vol. 30, no. 2, 263-275, 2000.

[36]   K. Huarng, and H.K. Yu, "A type 2 fuzzy time series model for stock index forecasting," *Physica A*, vol. 353, pp. 445-462, 2005.

[37]   H.K. Yu, "A refined fuzzy time series model for forecasting," *Physica A*, vol. 346, pp. 657-681, 2005.

[38]   H.K. Yu, "Weighted fuzzy time series models for TAIEX forecasting," *Physica A*, vol. 349, pp. 609-624, 2005.

[39]   S.M. Chen, "Forecasting enrollments based on high order fuzzy time series," *Cybernet Syst*, vol. 33, pp. 1-16, 2002.

[40]   C.H. Aladag, M.A. Basaran, E. Eğrioğlu, U. Yolcu, and V.R. Uslu, "Forecasting in high order fuzzy times series by using neural networks to define fuzzy relations," *Expert Syst Appl*, vol. 36, no. 3, pp. 4228–4231, 2009.

[41]   E. Eğrioğlu, C.H. Aladag, U. Yolcu, V.R. Uslu, M.A. Basaran, "A new approach based on artificial neural networks for high order multivariate fuzzy time series," *Expert Syst Appl*, vol. 36, no. 7, pp. 10589-10594, 2009.

[42]   T.K. Yu, and K. Huarng, "A bivariate fuzzy time series model to forecast the TAIEX," *Expert Syst Appl*, vol. 34, no. 4, pp. 2945-2952, 2008.

[43]   L.W. Lee, L.H. Wang, S.M. Chen, and Y.H. Leu, "Handling forecasting problems based on two factors high order fuzzy time series," *IEEE Trans Fuzzy Syst*, vol. 14, pp. 468-477, 2006.

[44]   T.A. Jilani, and S.M.A. "Burney, Multivariate stochastic fuzzy forecasting models," *Expert Syst Appl*, vol. 35, pp. 691-700, 2008.

[45]   P.F. Pai, and C.S. Lin, "A Hybrid ARIMA and support vector machines model in stock price forecasting," *Int J Manag Sci*, vol. 33, pp. 497-505, 2005.

[46]   F.M. Tseng, H.C. Yu, and G.H. Tzeng, "Combining neural network model with seasonal time series ARIMA model," *Technol Forecast Soc*, vol. 69, pp. 71-87, 2002.

[47]   G. Zhang, "Time series forecasting using a hybrid ARIMA and neural network model," *Neurocomputing*, vol. 50, pp. 159-175, 2003.

# CHAPTER 2

# Comparison of Feed Forward and Elman Neural Networks Forecasting Ability: Case Study for IMKB

## Erol Eğrioğlu[1,*], Cagdas Hakan Aladag[2] and Ufuk Yolcu[1]

*[1]Ondokuz Mayis University, Faculty of Arts and Science, Department of Statistics, 55139, Samsun, Turkey and [2]Hacettepe University, Faculty of Science, Department of Statistics, 06800, Ankara, Turkey*

**Abstract:** In recent years, artificial neural networks (ANN) have been widely used in real life time series forecasting. Artificial neural networks can model both linear and curvilinear structure in time series. Most of the conventional methods used in the analysis of time series are linear structure and fail to analyze non-linear time series. In conventional time series methods such as threshold autoregressive, bilinear model, which are used in non-linear time series modeling, a particular curvilinear model pattern is needed. Artificial neural network is a method based on data and does not require a model pattern. With its activation function, it provides flexible non-linear modeling. Additionally, when compared with conventional methods, successful results are obtained in forecasting time series *via* artificial neural networks in the literature. In this study, feed forward and feedback artificial neural networks which are widely used for time series forecasting were applied to Istanbul Stock Exchange Market (IMKB) time series and forecasting performances were evaluated.

**Keywords:** Artificial neural networks, Feed forward, Feedback, Forecasting, Time series.

## 1. INTRODUCTION

Box-Jenkins method has been widely used in forecasting linear time series [1]. In real life, time series generally have curvilinear structure [2]. Since curvilinear time series cannot be modeled *via* Box-Jenkins methods, curvilinear time series models such as Autoregressive Model with Changing Conditional Variance [3], Threshold Models [4] and Bilinear Models [5] are used. As curvilinear time series models are only used for particular curvilinear structures, they do not have enough flexibility [6].

Artificial neural networks have the ability to learn both curvilinear and linear structures in time series. Compared with other time series modeling methods, they provide better forecasting results. Therefore, over the last two decades, there have been studies in the literature concerning the analyses of time series using artificial neural networks [6]. It is crucial to determine the architecture structure, learning and training parameters, activation function and training algorithm which consist of components such as number of input, number of hidden layer, number of units in hidden layer and the number of output. Studies in the literature are conflicting and no common conclusion can be reached in determining these components. No artificial neural network model applicable to each problem can be identified. The simulation studies aiming to determine the components for the forecasting problem are limited and no general conclusions can be reached [7]. Therefore, all components should be reviewed for each question. Feed forward artificial neural networks and Elman feedback artificial neural networks have been widely used in the literature for time series forecasting [8-10].

In this study, forecasting performances of feed forward neural networks and Elman feedback artificial neural networks were discussed by analyzing application results of IMKB time series. Feed forward artificial neural networks and Elman feedback artificial neural networks were described in the second section. Stages of forecasting using artificial neural networks were given as algorithms in the third section. Fourth section included detailed results obtained from the analysis of IMKB time series *via* feed forward and Elman feedback artificial neural networks. Data obtained from the application were evaluated in the discussion section.

---

*Address correspondence to Erol Eğrioğlu:* Ondokuz Mayis University, Faculty of Arts and Science, Department of Statistics, 55139, Samsun, Turkey; E-mail: erole@omu.edu.tr

## 2. ARTIFICIAL NEURAL NETWORKS

Artificial neural network is a data processing mechanism generated by the simulation of human nerve cells and nervous system in a computer environment. The most important feature of artificial neural network is its ability to learn from the examples. Despite having a simpler structure in comparison with the human nervous system, artificial neural networks provide successful results in solving problems such as forecasting, pattern recognition and classification.

Although there are many types of artificial neural networks in the literature, feed forward and Elman feedback artificial neural networks are frequently used for forecasting problem. Feed forward artificial neural networks consist of input layer, hidden layer(s) and output layers. An example of feed forward artificial neural network architecture is shown in Fig. **1**. Each layer consists of units called neuron and there is no connection between neurons which belong to same layer. Neurons from different layers are connected to each other with their weights. Each weight is shown with directional arrows in Fig. **1**. Bindings shown with directional arrows in feed forward artificial neural networks are forward and unidirectional. In the literature, many studies on forecasting use single neuron in output layer. An activation function is used for each neuron in hidden layer and output layer of feed forward artificial neural networks. Inputs incoming to neurons in hidden and output layers are made up multiplication and addition of neuron outputs in the previous layers with the related weights. Data from these neurons pass through the activation function and neuron output are formed. Activation function enables curvilinear match-up. Therefore, non-linear activation functions are used for hidden layer units. In addition to a non-linear activation function, linear (pure linear) activation function can be used in output layer neuron.

In feed forward artificial neural networks, learning is the determination of weights generating the closest outputs to the target values that correspond with the inputs of artificial neural network. Learning is achieved by optimizing the total errors with respect to weights. There are several types of training algorithms in the literature used for learning of feed forward artificial neural networks. One of the widely used training algorithms is Levenberg-Marquardt (LM) [12] algorithm which was also used in this study.

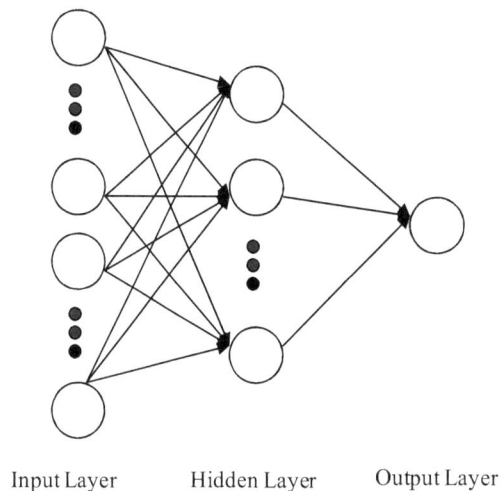

**Figure 1:** Multilayer feed forward artificial neural network with one output neuron.

Elman artificial neural network is one of the important artificial neural network type used in forecasting time series. Elman neural network, which has the simplest structure among feedback artificial neural networks types, was first proposed by Elman [11]. Elman feedback artificial neural networks consist of input layer, hidden layer, context layer and output layer. Context layer provides a step-delayed feedback mechanism which shows hidden layer output to network as input thus enabling artificial neural network learning with more information. An example of Elman artificial neural network architecture is shown in Fig. **2**.

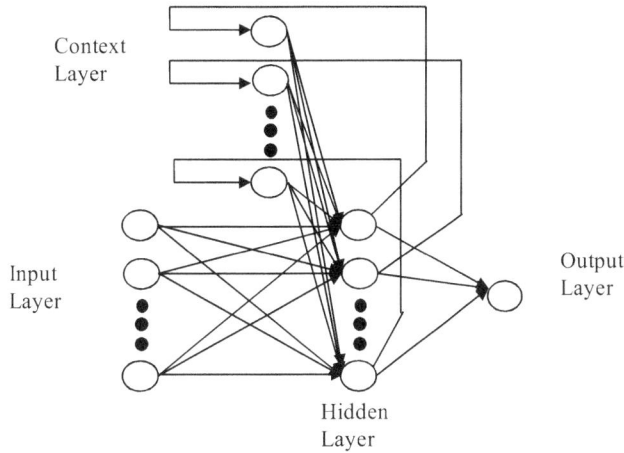

**Figure 2:** Elman recurrent artificial neural network.

## 3. FORECASTING TIME SERIES WITH ARTIFICIAL NEURAL NETWORKS

Recently, artificial neural networks have been used in most time series forecasting applications. The following factors lead the widespread usage of artificial neural networks method in time series:

- Analysis can be done without testing whether time series have curvilinear or linear structure.

- When compared with conventional time series method, artificial neural network method provides much better forecasting results.

- As conventional time series models can only be used for particular curvilinear structures, they are not flexible in general. But in the analysis using artificial neural network, no matter what the curvilinear structure of time series is.

- Artificial neural network theory is not complex as in the conventional time series forecasting methods and it is easy to understand.

The use of both feed forward artificial neural networks and feedback neural networks in forecasting time series can be summarized in seven steps.

***Step 1.*** Selection of activation function and pre-processing of the data.

Initially, the type of activation function to be used in hidden and output layer of artificial neural network is selected. Logistic activation function was used in hidden layer units in the application of this study. Logistic activation function is given as follows:

$$f(x) = \frac{1}{1 + \exp(-x)} \tag{1}$$

Data are converted into the range suitable for activation function. If the logistic activation function is to be used, *xi,* to show input value can be converted in to (0, 1) range like,

$$x_i' = \frac{x_i - Min(x_i)}{Maks(x_i) - Min(x_i)} \tag{2}$$

where, *Max $(x_i)$* represents maximum input value; *Min $(x_i)$* represents minimum input value.

***Step 2.*** Data organization

In this step, it is determined that how much percentage of the data set, training and data size constitutes.

***Step 3.*** Modeling

Artificial neural network model is established by determining the number of input, the number of hidden layer, the number of unit in hidden layer, the number of unit in output layer, activation function used in the unit, training algorithm and parameters of the algorithm and performance criteria.

***Step 4.*** Generating input values

Input values of artificial neural network are delayed time series. Let $m$, be the number of units in input layer while generating input value for $X_t$ time series. $m$ number of delayed time series are generated as $X_{t-1}$, $X_{t-2}$,..., $X_{t-m}$.

***Step 5.*** Calculation of the best weight values

The best weight values are found on training set *via* selected learning algorithm. Output values of established artificial neural model are calculated.

***Step 6.*** Calculation of the Performance criteria

Forecasts of artificial neural network test set are obtained. Inverse transformation in step 1 is applied to output values obtained in step 5 and values obtained in this step. Resultant values constitute forecasts of training set and test set respectively. Based on the differences between forecasts of test set and data in test set, selected performance criterion is calculated. There are several performance criteria used to determine that how well artificial neural network learn the relation in the data. One of the most commonly used performance criteria is the root of mean square error (RMSE). Let $z$ be the artificial neural network output value, $d_p$ target output value and $p$, the growth of test set, RMSE can be calculated with the following equation:

$$RMSE = \sqrt{\frac{\sum_{p=1}^{P}(d_p - z_p)^2}{p}}$$
(3)

***Step 7.*** Forecasting

Finally, forecast values are obtained for the future using the best weight values obtained in step 5.

## 4. THE IMPLEMENTATION

The proposed method is applied to the time series of IMKB 100 index which is between May 20, 2008 and September 29, 2008. The time series consist of 95 observations and its graph is shown in Fig. **3**. The first 80 observations are used for training and the last 15 observation are used for test set. A total of four models (two models based on feed forward artificial neural network and two based on Elman feedback artificial neural network) were used for the analysis of IMKB time series. These four models include following components;

**Model 1.** Feed forward artificial neural network with single hidden layer and single output unit in which logistic activation function is used in hidden layer units whereas linear activation function is used in output layer.

**Model 2.** Feed forward artificial neural network with single hidden layer and single output unit in which logistic activation functions is used in hidden layer units and output layer units.

**Model 3.** Elman feedback artificial neural network with single hidden layer and single output unit in which logistic activation function is used in hidden layer units whereas linear activation function is used in output layer.

**Model 4.** Elman feedback artificial neural network with single hidden layer and single output unit in which logistic activation functions is used in hidden layer units and output layer units.

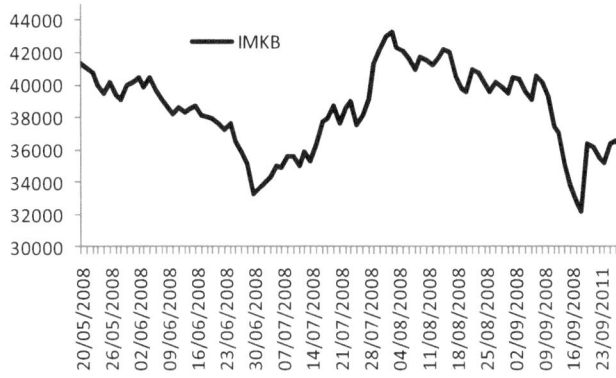

**Figure 3:** The time series data of Istanbul IMKB Index 100.

In the application of these models in IMKB 100 Index time series, a total of 144 different analyses were made by varying input unit number and hidden layer unit number from 1 to 12 for each model. In each artificial neural network training, LM algorithm is applied considering the iteration number as 100. RMSE values calculated for test set from the model analyses are given in Tables **1-4**.

**Table 1:** The obtained results of Model 1

| Number of hidden layer Neurons | Number of input layer neurons | | | | | | | | | | | |
|---|---|---|---|---|---|---|---|---|---|---|---|---|
| | 1 | 2 | 3 | 4 | 5 | 6 | 7 | 8 | 9 | 10 | 11 | 12 |
| 1 | 1381 | 1354 | 1355 | 1342 | 1398 | 1368 | 1379 | 1425 | 1362 | 1361 | 1372 | 1388 |
| 2 | 1521 | 1275 | 1355 | 1390 | 1580 | 4988 | 1439 | 293923 | 65564 | 1226 | 1327 | 432473 |
| 3 | 1327 | 1455 | 5968 | 1535 | 3830 | 141876 | 1766 | 2358 | 6539 | 34640 | 1253 | 10035 |
| 4 | 1266 | 1300 | 1649 | 1981 | 2186 | 137869 | 150870 | 3498 | 2303 | 40289 | 7148 | 25065 |
| 5 | 1354 | 1410 | 2944 | 2464 | 41120 | 247376 | 47496 | 1684 | 7391 | 11053 | 7217 | 5122 |
| 6 | 1387 | 1593 | 49534 | 6144 | 1756 | 92873 | 2483 | 88068 | 28849 | 11476 | 5545 | 10474 |
| 7 | 1391 | 1865 | 20069 | 25634 | 20272 | 5509 | 6128 | 13508 | 16674 | 7098 | 9610 | 5988 |
| 8 | 4962 | 36213 | 6991 | 8799 | 3196 | 7249 | 32200 | 5683 | 6717 | 11669 | 2805 | 2641 |
| 9 | 2868 | 94069 | 234166 | 44285 | 27580 | 48440 | 19593 | 12437 | 3197 | 4156 | 3428 | 3336 |
| 10 | 2780 | 33234 | 66734 | 19468 | 22301 | 7784 | 14753 | 6282 | 9845 | 5813 | 3946 | 3323 |
| 11 | 16809 | 7545 | 10023 | 11111 | 31007 | 17336 | 10600 | 3317 | 4383 | 5917 | 3517 | 4547 |
| 12 | 1312 | 2303 | 9155 | 23017 | 14684 | 30445 | 11576 | 10429 | 5565 | 4861 | 3203 | 5478 |

**Table 2:** The obtained results of Model 2

| Number of hidden layer Neurons | Number of input layer neurons | | | | | | | | | | | |
|---|---|---|---|---|---|---|---|---|---|---|---|---|
| | 1 | 2 | 3 | 4 | 5 | 6 | 7 | 8 | 9 | 10 | 11 | 12 |
| 1 | 1303 | 1278 | 1278 | 1287 | 1304 | 1300 | 1345 | 1379 | 1317 | 1323 | 1335 | 1354 |
| 2 | 7561 | 1278 | 7561 | 1373 | 1360 | 1625 | 1591 | 1685 | 1755 | 1323 | 1388 | 1643 |
| 3 | 1288 | 1376 | 1330 | 1396 | 1979 | 2422 | 2248 | 2851 | 1682 | 1379 | 3705 | 1791 |

*Table 2: cont….*

| | | | | | | | | | | | | |
|---|---|---|---|---|---|---|---|---|---|---|---|---|
| 4 | 1266 | 1254 | 1683 | 4019 | 2490 | 1925 | 2182 | 1881 | 2617 | 2064 | 2344 | 1879 |
| 5 | 1354 | 1428 | 2224 | 1818 | 3698 | 2315 | 3295 | 3256 | 3327 | 5245 | 4642 | 4045 |
| 6 | 1349 | 1426 | 2574 | 1477 | 3181 | 1780 | 2679 | 3496 | 3659 | 2290 | 4254 | 2968 |
| 7 | 1435 | 4114 | 7561 | 2067 | 4237 | 2831 | 5103 | 3273 | 5683 | 4365 | 4652 | 2109 |
| 8 | 1333 | 1661 | 4560 | 2649 | 4230 | 4185 | 6595 | 4490 | 3131 | 3131 | 4268 | 4157 |
| 9 | 1960 | 2840 | 4606 | 7561 | 4065 | 6155 | 4074 | 4361 | 2430 | 2487 | 2768 | 2960 |
| 10 | 1602 | 3559 | 2054 | 2539 | 3776 | 4894 | 4755 | 3693 | 3009 | 3058 | 2108 | 2666 |
| 11 | 1300 | 3865 | 4173 | 2867 | 5083 | 4602 | 3451 | 5731 | 4761 | 3333 | 2913 | 2889 |
| 12 | 1625 | 2174 | 5320 | 3064 | 4343 | 3684 | 4448 | 4134 | 3018 | 2686 | 7561 | 3398 |

**Table 3:** The obtained results of Model 3

| Number of hidden layer Neurons | Number of input layer neurons | | | | | | | | | | | |
|---|---|---|---|---|---|---|---|---|---|---|---|---|
| | 1 | 2 | 3 | 4 | 5 | 6 | 7 | 8 | 9 | 10 | 11 | 12 |
| 1 | 1381 | 1354 | 1355 | 1342 | 1398 | 1368 | 1379 | 1425 | 1362 | 1361 | 1372 | 1388 |
| 2 | 1520 | 1354 | 5989 | 1314 | 1225 | 4999 | 4024 | 1529 | 1362 | 63110 | 1327 | 456498 |
| 3 | 1589 | 1455 | 1703 | 1393 | 1709 | 1370 | 50442 | 1662 | 1463 | 1631 | 6545 | 2347 |
| 4 | 1149 | 1560 | 13883 | 1986 | 3372 | 2312 | 278698 | 26725 | 14121 | 4589 | 41550 | 4825 |
| 5 | 1211 | 1713 | 3369 | 2928 | 85141 | 2063 | 2075 | 221826 | 8363 | 38789 | 6553 | 6450 |
| 6 | 1298 | 2696 | 2663 | 168180 | 38288 | 4017 | 6269 | 17781 | 13840 | 5323 | 4770 | 7333 |
| 7 | 1433 | 1961 | 5391 | 166315 | 129390 | 2905 | 5153 | 11041 | 10848 | 8427 | 5308 | 3644 |
| 8 | 1616 | 64230 | 5235 | 37316 | 5670 | 12954 | 22469 | 29263 | 5053 | 4258 | 5766 | 5281 |
| 9 | 1256 | 18882 | 9096 | 7159 | 6393 | 43993 | 9676 | 14571 | 6236 | 4912 | 17789 | 4221 |
| 10 | 1943 | 26081 | 51697 | 33975 | 13596 | 12359 | 14790 | 3710 | 10705 | 3563 | 2835 | 5721 |
| 11 | 1483 | 2908 | 8422 | 25725 | 17231 | 16504 | 29022 | 7191 | 11801 | 8473 | 5259 | 3661 |
| 12 | 1156 | 4372 | 5715 | 5312 | 16212 | 14454 | 11153 | 12360 | 10513 | 5894 | 3169 | 2790 |

**Table 4:** The obtained results of Model 4

| Number of hidden layer Neurons | Number of input layer neurons | | | | | | | | | | | |
|---|---|---|---|---|---|---|---|---|---|---|---|---|
| | 1 | 2 | 3 | 4 | 5 | 6 | 7 | 8 | 9 | 10 | 11 | 12 |
| 1 | 1303 | 1278 | 1278 | 1286 | 1304 | 1301 | 1345 | 3501 | 1317 | 1323 | 1335 | 1354 |
| 2 | 1303 | 1288 | 1298 | 1373 | 1305 | 1513 | 1409 | 1731 | 1340 | 1268 | 1335 | 1690 |
| 3 | 1263 | 1376 | 1718 | 1701 | 3976 | 1935 | 2299 | 1378 | 2272 | 1434 | 4432 | 1426 |
| 4 | 1311 | 1347 | 2140 | 3623 | 3065 | 2361 | 2807 | 3543 | 2699 | 3781 | 3152 | 3091 |
| 5 | 1406 | 7561 | 7561 | 2834 | 1749 | 7307 | 2789 | 2168 | 4253 | 3407 | 3226 | 3742 |
| 6 | 1227 | 2005 | 3594 | 1838 | 4113 | 3447 | 3747 | 3926 | 4311 | 4226 | 7561 | 4472 |
| 7 | 1386 | 1374 | 2898 | 3301 | 4873 | 3239 | 4141 | 4475 | 5325 | 3925 | 2663 | 3530 |
| 8 | 1261 | 2904 | 3993 | 4096 | 3369 | 5073 | 3492 | 4709 | 3472 | 4435 | 5021 | 3542 |
| 9 | 1298 | 2499 | 3901 | 1924 | 3863 | 5881 | 5875 | 3823 | 4852 | 3551 | 4976 | 2909 |
| 10 | 1303 | 1986 | 6122 | 4193 | 4261 | 3488 | 3721 | 5758 | 3728 | 3640 | 3150 | 3283 |
| 11 | 1340 | 4292 | 2370 | 4855 | 7561 | 4962 | 4530 | 3450 | 2073 | 2071 | 3717 | 3054 |
| 12 | 1511 | 4162 | 3724 | 3312 | 7561 | 4036 | 4440 | 3017 | 3666 | 3115 | 3768 | 5322 |

When the results in Tables **1-4** were analyzed, it was seen that the best results for model 1 were achieved when input layer unit number was 10 and hidden layer unit number was 2 and RMSE value was 1226. For model 2, the best results were achieved when input layer unit number was 2, hidden layer unit number was 4 and RMSE value was 1254. For model 3, the best results were achieved when input layer unit number was 1, hidden layer unit number was 6 and RMSE value was 1149. For model 4, the best results were achieved when input layer unit number was 1, hidden layer unit number was 6 and RMSE value was 1227. Consequently, the most accurate results were obtained when model 3 with 1 input and 6 units in the hidden layer was used.

## 5. CONCLUSION

Real life time series may include both linear and non-linear structures. Elman feedback artificial neural network is one of the most effective methods used in modeling of non-linear time series. In this study, IMKB 100 Index time series were analyzed using both feed forward and Elman feedback artificial neural networks. At the end of the application, Elman feedback artificial neural network with single neuron in input layer, 4 neurons in hidden layer and single unit in single output layer and in which logistic activation model was used in hidden layer units and linear activation function was used in output layer was found the best artificial neural network for IMKB 100 Index. Graph of the forecasts obtained from this model and real values are given in Fig. **4**. According to Fig. **4**, it can be said that the calculated forecasts over the test set for this model is very good.

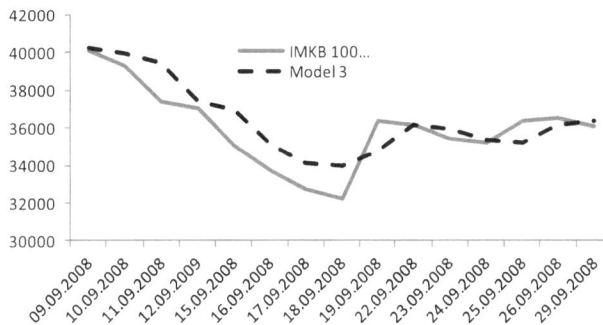

**Figure 4:** The time series test data of Istanbul IMKB Index 100 and the best results of Model 3.

## REFERENCES

[1]     G.E.P. Box, and G.M. Jenkins, *Time Series Analysis: forecasting and control*. San Francisco: Holden Day, 1976.

[2]     C.W.J. Granger, and T. Terasvirta, *Modelling Nonlinear Economic Relationships*. Oxford University Press: Oxford, 1993.

[3]     R.F. Engel, "Autoregressive Conditional Heteroskedasticity with Estimates of the variance of UK Inflation," *Econometrica*, vol. 50, pp. 987-1008, 1982.

[4]     C.H. Aladag, E. Eğrioğlu, S. Gunay and M.A. Basaran, "Improving weighted information criterion by using optimization," *J Comput Appl Math*, vol. 233, pp. 2683-2687, 2010.

[5]     C.W.J. Granger, and A.P. Anderson, *An Introduction to Bilinear Time Series Models.* Vandenhoeck and Ruprecat: Göttingen, 1978.

[6]     G. Zhang, B.E. Patuwo, and Y.M. Hu, "Forecasting with artificial neural networks: the state of the art," *Int J Forecasting*, vol. 14, pp. 35-62, 1998.

[7]     H.B. Hwrang, "Insights into Neural Network Forecasting of the time Series Corresponding to ARMA(p,q) Structures," *Int J Manag Sci*, vol. 29, pp. 273-289, 2001.

[8]     C.H. Aladag, E. Eğrioğlu, and C. Kadilar, "Forecasting nonlinear time series with a hybrid methodology, *Appl Math Lett*," vol. 22, pp. 1467-1470, 2009.

[9]     C.H. Aladag, E. Eğrioğlu, and S. Gunay, "A new architecture selection strategy in solving seasonal autoregressive time series by artificial neural networks," *Hacet J Math Stat*, vol. 37, no. 2, pp. 185-200, 2008.

[10]    E. Eğrioğlu, C.H. Aladag, and S. Gunay, "A new model selection strategy in artificial neural network," *Appl Math Comput*, vol. 195, pp. 591-597, 2008.

[11]    J.L. Elman, "Finding structure in time," *Cognitive Science*, vol. 14, pp. 179-211, 1990.

[12]    K. Levenberg, "A method for the solution of certain non-linear problems in least squares," *Q Appl Math*, vol. 2, pp. 164–168, 1944.

*Advances in Time Series Forecasting*, 2012, 18-25

# CHAPTER 3

# Comparison of Architecture Selection Criteria in Analyzing Long Memory Time Series

## Erol Eğrioğlu[1], Cagdas Hakan Aladag[2,*] and Ufuk Yolcu[1]

*[1]Ondokuz Mayis University, Faculty of Arts and Science, Department of Statistics, 55139, Samsun, Turkey and [2]Hacettepe University, Faculty of Science, Department of Statistics, 06800, Ankara, Turkey*

**Abstract:** In recent years, studies including long memory time series are existed in the literature. Such time series in real life may have both linear and nonlinear structures. Linear models are inadequate for this kind of time series. An alternative method to forecast these time series is artificial neural networks which is data based and can model both linear and nonlinear structure in these time series. In order to determine the number of nodes in the layers of a network is an important decision. This decision has been made by using various architecture selection criteria. The performance of these criteria varies, depending on components of time series, such as trend and seasonality. In this study, some architecture selection criteria are compared on real time series when artificial neural networks are employed in forecasting. Some advices are given for using artificial neural networks to forecast long memory time series.

**Keywords:** Architecture selection criteria, Artificial neural networks, Long range dependent, Time series.

## 1. INTRODUCTION

In real life, many time series can have long memory structure. These time series can be called long range dependent or long memory time series. In the literature, Autoregressive Fractionally Integrated Moving Average (ARFIMA) models have been used to analyze these time series. Despite most of the long range dependent time series faced in real life can have a non linear structure, ARFIMA models are linear. Therefore, using ARFIMA models for these time series can lead to misleading results. On the other hand, using artificial neural networks would be a wiser choice since they can model both linear and non linear structures in time series.

When time series are forecasted by using artificial neural networks, some criteria such as root mean square error (RMSE), mean absolute percent error (MAPE), Akaike information criterion (AIC), Bayesian information criterion (BIC), direction accuracy criterion (DAC) are employed for architecture selection. Qi and Zhang [1] studied on using these selection criteria for time series which does not include long memory structure. In their study, they found that the mentioned selection criteria produce inconsistent results for in and out sample observations. In other words, there is no guarantee that the architecture found as the best architecture for in sample observations is the best architecture for out sample observations. Qi and Zhang [1] also pointed out that AIC and BIC criteria produce very poor results when artificial neural networks are used for forecasting.

In this study, we aim to examine whether the results reached by Qi and Zhang [1] are valid for long range dependent time series or not. In addition, it is aimed to bring out which criterion give best results for these time series. For doing these, yearly minimal water levels of the Nile River [2], VBR data [2], and Nasqad-100 index volatile [3] long range dependent time series are used. In addition to these series, the international tourism demand of Turkey time series, which has long memory structure, is used in the implementation. Besides, some time series component such as seasonality and trend can affect the performance of the selection criteria. Therefore, it should be examined that whether long range dependency has an effect on the performance of the criteria or not.

The next section introduces the long range dependent time series. In Section 3, brief information about artificial neural networks is given. Section 4 presents the architecture selection criteria such as RMSE,

---

*\*Address correspondence to Cagdas Hakan Aladag:* Hacettepe University, Faculty of Science, Department of Statistics, 06800, Ankara, Turkey; E-mail: chaladag@gmail.com

MAPE, AIC, BIC, and DAC. While the long range dependent time series are being forecasted with artificial neural networks, performances of mentioned criteria are examined in Section 5. The obtained results are discussed in the last section.

## 2. THE LONG RANGE DEPENDENT TIME SERIES

Long range dependent time series can be faced in many fields such as economy, hydrology, and geophysics. Existence of long range dependent was firstly introduced by Hurst [4]. In his study, he tried to determine the minimum level of the Nile River by utilizing the concept of long range dependent [4]. Therefore, the long range dependent can be called as Hurst effect in the literature. The long range dependent time series has some characteristics given below [2]:

i)     There are relatively long periods where the observations tend to stay at a high level, and on the other hand, there are long periods with low levels.

ii)    When short time periods are examined, local trends and periodical behaviors are observed. However, local trends or periodical behaviors are not clearly seen when the time series data are examined.

iii)   Overall, the series looks stationary.

In long range dependent process, some numerical properties can be given as follows [2]:

i)     The variance of the sample mean seems to decay to zero at a slower rate than $n^{-1}$. In good approximation, the rate is proportional to $n^{-\alpha}$, $0 < \alpha < 1$.

ii)    The sample correlations decay to zero at a rate that is in good approximation proportional to $k^{-\alpha}$, $0 < \alpha < 1$.

iii)   $I(\lambda)$ Near the origin, the logarithm of the periodogram plotted against the logarithm of the frequency appears to be randomly scattered around a straight line with negative slope.

Due to these given characteristics, the definition of long range dependency can be given as follows:

Let $\alpha \in (0,1)$ is a real number and $c_\rho > 0$ is a constant. If a stationary process $X_t$ satisfies the condition which can be given as follows:

$$\lim_{k \to \infty} \rho(k) / \left[ c_\rho k^{-\alpha} \right] = 1, \tag{1}$$

process $X_t$ can be called as long range dependent process. The detailed information about long range dependent time series can be found in [5].

## 3. ARTIFICIAL NEURAL NETWORKS

Artificial neural networks approach has been used as a good alternative method in time series analysis. Zhang [6] summarized some time series applications in which artificial neural networks are used as analyzing method. Zhang [6] revealed basic concepts of using artificial neural networks for time series analysis. Artificial neural networks approach is a data based approach. For the examined data, therefore, using right components of the method can lead better results. Selection of the components such as architecture structure, learning algorithm and activation function has an important effect on the performance of artificial neural networks [7]. The basic elements of ANN can be given as follows [8]:

*Architecture structure*: Feed forward ANN has been widely used for forecasting problems because of their simple usage and success. The structure of multilayer feed forward ANN is basically given in Fig. **1**.

Multilayer feed forward ANN as illustrated in the figure consist of three parts such as input, hidden, and output layers. Each layer consists of neurons. The architecture structure is determined based on deciding the number of neuron in each layer. These neurons are linked each other by weights. There is no link among the neurons in the same layer. For a forecasting problem, the inputs of the network are past lagged observations.

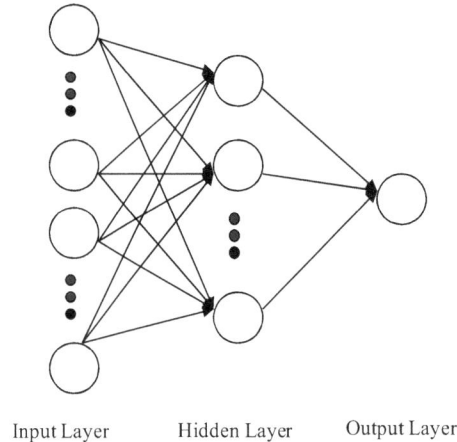

                 Input Layer        Hidden Layer        Output Layer

**Figure 1:** Multilayer feed forward ANN with one output neuron.

One critical decision is to determine the appropriate architecture, that is, the number of layers, number of nodes in each layer [9]. However, in the literature, there are not general rules for determining the best architecture [10].

*Learning algorithm*: There have been many learning algorithms in order to determine weights. The one of the most employed algorithm is called Back Propagation Learning Algorithm. This learning algorithm updates the weights based on difference between real value and output value of the ANN. However, back propagation networks have some disadvantages mentioned in the introduction. In light of the weakness of the conventional back propagation algorithm, a number of variations or modifications of this algorithm, such as the adaptive method, quickprop, and second-order methods *etc.*, have been proposed [11]. Among them, the second-order methods such as Levenberg Marquardt method are more efficient nonlinear optimization methods and are used in most optimization packages. Their faster convergence, robustness, and the ability to find good local minima make them attractive in ANN training [11]. Therefore, Levenberg Marquardt method [12] is used as training algorithm in the implementation.

*Activation function*: Activation function provides the non-linear mapping between input and output. The performance of networks depends on the proper choice of activation function. Activation function can be chosen as either linear or double polarized, or one polarized. Slope parameter should be determined when the activation is non linear. Also, slope parameter plays a key role in reaching desired output values.

In the literature, simulation studies conducted to determine the components given above are not enough since these studies are very limited and do not include general results [13]. Lots of component combinations have to be tried to determine the best components which gives the best results. Despite this disadvantage of artificial neural networks, the method has some advantages given below.

- Time series can be analyzed by artificial neural networks without testing time series for nonlinearity.

- Artificial neural networks can produce better results those obtained from conventional methods.

- Artificial neural networks can model time series that include any type of nonlinearity. On the other hand, other non linear time series methods can only model time series that have specific non linear structures since these methods were proposed for only specific structures.

- Artificial neural networks do not have complex theory unlike other conventional methods. Therefore, it is easier to use artificial neural networks in time series analysis.

The detailed information about the usage of artificial neural networks in time series analysis can be found in [14-17].

## 4. MODEL SELECTION CRITERIA

In the literature, there have been some methods for architecture selection in artificial neural networks. One of these methods is using a criterion such as RMSE and MAPE. Qi and Zhang [1] examined different performance criteria in artificial neural networks. In their study, however, performance criteria were not examined for long range dependent time series. The performance criteria, which are AIC, BIC, RMSE, MAPE, and DAC, examined by Qi and Zhang [1] can be calculated by using formulas given below.

$$AIC = \log\left(\frac{\sum_{i=1}^{T}(y_i - \hat{y}_i)^2}{T}\right) + \frac{2m}{T} \tag{2}$$

$$BIC = \log\left(\frac{\sum_{i=1}^{T}(y_i - \hat{y}_i)^2}{T}\right) + \frac{m\log(T)}{T} \tag{3}$$

$$RMSE = \left(\frac{\sum_{i=1}^{T}(y_i - \hat{y}_i)^2}{T}\right)^{1/2} \tag{4}$$

$$MAPE = \frac{1}{T}\sum_{i=1}^{T}\left|\frac{y_i - \hat{y}_i}{y_i}\right| \tag{5}$$

$$DAC = \frac{1}{T}\sum_{i=1}^{T}a_i \quad, \quad a_i = \begin{cases} 1 & if \ (y_{i+1} - y_i)(\hat{y}_{i+1} - y_i) > 0 \\ 0 & d.d. \end{cases} \tag{6}$$

In here, $y_i$, $\hat{y}_i$, $T$ and $m$ represent real values, forecasts, the number of observations and the number of parameter, respectively. In a model of artificial neural networks, parameters are weights included by the model.

In the literature, RMSE and MAPE criteria have been generally preferred for architecture selection. The architectures determined by using these two criteria can give forecasts which are very close to each others. So such results do not have any meaning in most of the applications. In addition, the architectures determined by RMSE or MAPE cannot model the turning points and this is a very vital problem. RMSE and MAPE do not penalize the number of parameter unlike AIC and BIC criteria. Criteria such as AIC and BIC are not good enough for the architecture selection since they are generally chose the architectures that have minimum parameters [1]. In [1], it also was revealed that DAC is not good enough for the architecture selection when only DAC is used to find the best architecture. Qi and Zhang [1] claimed that RMSE or MAPE can be used to determine the best architecture while the criteria such as AIC, BIC and DAC are not good enough for the architecture selection. The criteria like AIC, BIC and DAC can be employed to get extra information.

## 5. THE PERFORMANCES OF THE CRITERIA FOR LONG RANGE DEPENDENT TIME SERIES

The results obtained by Qi and Zhang [1] were given in the previous section. One of the purposes of this study is to bring out whether the results obtained by Qi and Zhang [1] are valid for long range dependent time series or not. Other purpose is to examine the best criterion or criteria. Thus some suggestions can be given based on the obtained results. For doing these, yearly minimal water levels of the Nile River [2] (NR), VBR data [2] (VBR), and Nasqad-100 index volatile (N100) [3] long range dependent time series are used. In addition, the international tourism demand of Turkey (TDT) time series is used since according to R/S test introduced in [18], it is seen that first order seasonal difference of this time series is a long range dependent time series.

In this study, NR, VBR, N100, and TDT time series are analyzed by using artificial neural networks. First of all time series is split into two sets and the first set is used for training. Then, the other set is divided into two subsets. For example, a time series containing 43 observations are split into two sets, including 35 and 8 observations, respectively. Thus training set consists of first 35 observations. Then, the set contains last 8 observations can be partitioned into two subsets. The first set can contain 5 and the second one can contain 3 observations. The first and second test sets are called validation and test sets, respectively. The given example is illustrated in Fig. **2** in which numbers represents the observation numbers. RMSE, MAPE, DAC, AIC and, BIC which are obtained from different artificial neural network models, are calculated for each two test sets which are validation and test sets.

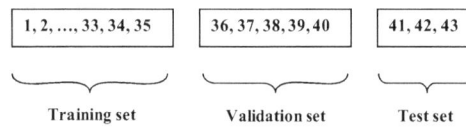

| 1, 2, ..., 33, 34, 35 | 36, 37, 38, 39, 40 | 41, 42, 43 |

Training set     Validation set     Test set

**Figure 2:** Data sets.

For NR time series contains 663 observations, the numbers of observations in the validation and the test sets are 60 and 63, respectively. For VBR contains 1000 observations, 100 and 100, for N100 contains 1456 observations, 140 and 140, for TDT contains 264 observations, 10 and 14 observations are used in the validation and the test sets, respectively.

Both the numbers of neurons in the input and the hidden layers are changed between 1 and 12. Thus 144 architectures are totally examined. For each architectures, all criteria values are calculated over both the validation and the test sets. Then, the calculated values are mapped into interval (0, 1). For instance, when N100 time series is analyzed, criteria values calculated over the validation set is given in Table **1**. In Table **1**, criteria values for only 12 architectures are presented as an example.

**Table 1:** Criteria values calculated over the validation set for N100

| *in* | *hn* | RMSE | MAPE | AIC | BIC | DAC |
|------|------|------|------|-----|-----|-----|
| 1 | 1 | 0.223 | 0.173 | 0.011 | 0.000 | 0.371 |
| 1 | 2 | 0.333 | 0.282 | 0.050 | 0.024 | 0.407 |
| 1 | 3 | 0.363 | 0.314 | 0.069 | 0.040 | 0.421 |
| 1 | 4 | 0.380 | 0.333 | 0.085 | 0.055 | 0.436 |
| 1 | 5 | 0.400 | 0.356 | 0.102 | 0.069 | 0.436 |
| 1 | 6 | 0.425 | 0.385 | 0.120 | 0.084 | 0.436 |
| 1 | 7 | 0.442 | 0.404 | 0.136 | 0.099 | 0.436 |
| 1 | 8 | 0.450 | 0.412 | 0.150 | 0.112 | 0.436 |
| 1 | 9 | 0.447 | 0.409 | 0.161 | 0.125 | 0.436 |
| 1 | 10 | 0.445 | 0.405 | 0.172 | 0.137 | 0.436 |

*Table 1: cont….*

| 1 | 11 | 0.446 | 0.407 | 0.184 | 0.150 | 0.436 |
| 1 | 12 | 0.446 | 0.408 | 0.196 | 0.162 | 0.436 |

In Table **1**, all criteria values for 12 architectures are shown. In the table, *in* and *hn* represent the number of neurons in the input and the hidden layers, respectively. For example, when the number of neurons in the input and the hidden layers are 1 and 4, respectively, the calculated MAPE value is 0.333. For these architectures, all criteria values calculated over the test set are presented in Table **2**.

**Table 2:** Criteria values calculated over the test set for N100

| *in* | *hn* | RMSE | MAPE | AIC | BIC | DAC |
|------|------|------|------|-----|-----|-----|
| 1 | 1 | 0.272 | 0.319 | 0.000 | 0.000 | 0.436 |
| 1 | 2 | 0.423 | 0.445 | 0.039 | 0.024 | 0.450 |
| 1 | 3 | 0.466 | 0.484 | 0.059 | 0.040 | 0.450 |
| 1 | 4 | 0.489 | 0.506 | 0.075 | 0.054 | 0.450 |
| 1 | 5 | 0.517 | 0.531 | 0.092 | 0.069 | 0.450 |
| 1 | 6 | 0.550 | 0.561 | 0.110 | 0.084 | 0.450 |
| 1 | 7 | 0.572 | 0.582 | 0.127 | 0.098 | 0.450 |
| 1 | 8 | 0.583 | 0.592 | 0.141 | 0.112 | 0.450 |
| 1 | 9 | 0.578 | 0.589 | 0.152 | 0.124 | 0.450 |
| 1 | 10 | 0.574 | 0.585 | 0.164 | 0.136 | 0.450 |
| 1 | 11 | 0.575 | 0.587 | 0.177 | 0.149 | 0.450 |
| 1 | 12 | 0.576 | 0.588 | 0.189 | 0.162 | 0.450 |

A good criterion should be consistent. In other words, a good criterion should produce consistent results for the validation and the test sets. To measure the consistency of a criterion, the correlation coefficient for the criterion values obtained over the validation and test sets should be calculated. The higher the correlation coefficient value between the validation and test sets is, the more consistent the criterion is. Therefore, as a result of the implementation, for all criteria, calculated correlation coefficients for all time series are summarized in Table **3**. In the implementation, for both the validation and test sets, 144 criterion values are obtained since 144 architectures is examined for each time series.

**Table 3:** Calculated correlation coefficients

|  | Time series | | | |
|------|------|------|------|------|
|  | TDT | VBR | NR | N100 |
| **RMSE** | 0.291 | -0.293 | 0.062 | 0.730 |
| **MAPE** | 0.255 | 0.938 | 0.313 | 0.850 |
| **AIC** | 0.990 | 0.700 | 0.990 | 0.990 |
| **BIC** | 0.990 | 0.940 | 0.990 | 0.990 |
| **DAC** | 0.081 | 0.535 | -0.308 | 0.810 |

First of all, we would like to note that correlation coefficients for AIC and BIC will always be very high since these criteria always take high values for bigger architectures and take less values for smaller architectures [7]. It was already mentioned that using a criterion like AIC, BIC and DAC for architecture selection is not a wise choice. According to Table **3**, for TDT, correlation coefficient value of RMSE is greater than correlation value of MAPE. On the other hand, both values are very small and indicate a poor correlation. Thus it cannot be said that these criteria are consistent for TDT time series. For VBR, it is obvious that MAPE is more consistent than RMSE. The obtained results for MAPE are very consistent and reliable. RMSE is inconsistent criterion for this time series. Misleading results will be obtained if RMSE criterion is used for this time series. For NR, MAPE looks more consistent than RMSE but the correlation coefficient is not high enough to get reliable results. Finally, for N100, it can be clearly said that both criteria are consistent and reliable. However, using MAPE instead of RMSE will be wiser since correlation coefficient value calculated for MAPE is greater.

To sum up, it cannot be said that whether MAPE is more consistent than RMSE or not according to obtained results. However, it is clearly seen that these criteria do not have certain behaviors when these criteria are used for long range dependent time series.

## 6. RESULTS AND DISCUSSION

Using artificial neural networks to analyze time series provides some important advantages. In the same way, when long range dependent time series are analyzed, utilizing artificial neural networks can lead to more accurate results than those obtained from other conventional methods. However, there are some problems with using artificial neural networks. One of these problems is to determine the best architecture which gives the best results. In the literature, some criteria have been used to solve architecture selection problem. Some of these criteria are RMSE, MAPE, DAC, AIC and, BIC. Qi and Zhang [1] examine these selection criteria for time series which do not have long range dependency.

It is a fact that some time series component such as seasonality and trend can affect the performance of the selection criteria. Therefore, long range dependency should also be examined for selection criteria. In the implementation, NR, VBR, N100, and TDT long range dependent time series are analyzed by using artificial neural networks. And RMSE, MAPE, DAC, AIC and, BIC criteria are employed to determine best architecture. Qi and Zhang [1] pointed out that these criteria give inconsistent results for time series which does not include long range dependency. In this study, correlation coefficients are utilized in order to examine the performance of the selection criteria. The long range dependent time series time series are divided into three sets. And by using last two sets, which are called the validation and test sets, these coefficients values are calculated. As a result of the implementation, it is seen that general results for these criteria are not obtained when long range dependent time series are analyzed with artificial neural networks.

## REFERENCES

[1]   M. Qi, and G. Zhang, "An investigation of model selection criteria for neural network time series forecasting," *Eur J Oper Res*, vol. 132, pp. 666-680, 2001.
[2]   J. Beran, *Statistics for long-memory processes*. Chapman&Hall/Crc, 1994.
[3]   E. Zivot, and D. Wang, *Modeling Financial Time Series with S-Plus*. Springer, New York, 2006.
[4]   H.E. Hurst, "Long-term storage capacity of reservoirs," *Trans. Am. Soc. Civil Engineers*, vol. 116, pp. 770-799, 1951.
[5]   E. Eğrioğlu, "Bayesian approaches on autoregressive fractionally integrated moving average models", Ph.D. thesis, Hacettepe University, Turkey, 2006.
[6]   G. Zhang, B.E. Patuwo, and Y.M. Hu, "Forecasting with artificial neural networks: the state of the art," *Int J Forecasting*, vol. 14, pp. 35-62, 1998.
[7]   C.H. Aladag, E. Eğrioğlu, S. Gunay, and M.A. Basaran, "Improving weighted information criterion by using optimization," *J Comput Appl Math*, vol. 233, pp. 2683-2687, 2010.
[8]   E. Eğrioğlu, C.H. Aladag, U. Yolcu, V.R. Uslu, and M.A. Basaran, "A new approach based on artificial neural networks for high order multivariate fuzzy time series," *Expert Syst Appl*, vol. 36, no. 7, pp. 10589-10594, 2009.
[9]   J.M. Zurada, *Introduction of Artificial Neural Systems*. St. Paul: West Publishing, 1992.
[10]  C.H. Aladag, "A new architecture selection method based on tabu search for artificial neural networks," *Expert Syst Appl*, vol. 38, pp. 3287–3293, 2011.
[11]  H.F. Zou, G.P. Xia, F.T. Yang, and H.Y. Wang, "An investigation and comparison of artificial neural network and time series models for Chinese food grain price forecasting," *Neurocomputing*, vol. 70, pp. 2913-2923, 2007.
[12]  K. Levenberg, "A method for the solution of certain non-linear problems in least squares," *Q Appl Math*, vol. 2, pp. 164–168, 1944.

[13]  H.B. Hwrang, "Insights into Neural Network Forecasting of the time Series Corresponding to ARMA(p,q) Structures," *Int J Manag Sci*, vol. 29, pp. 273-289, 2001.

[14]  E. Eğrioğlu, and C.H. Aladag, "A hybrid neural network and ARIMA model for forecasting in time series," In: *7th Turkey Econometrics and Statistics Symposium*, Malatya, Turkey, 2005. (http://www.ekonometridernegi.org/bildiriler/bildiriler.htm)

[15]  C.H. Aladag, and E. Eğrioğlu, "An application for forecasting in time series with using neural networks," In: *14th Statistics Research Symposium Proceedings*, 2005, pp. 397-406.

[16]  C.H. Aladag, and E. Eğrioğlu, "A simulation study to define a general model for ARMA structured time series," In: *5th Statistics Days Symposium Proceedings*, 2006, pp. 397-406.

[17]  E. Eğrioğlu, and C.H. Aladag, "Time series forecasting using a hybrid ARIMA and Elman neural networks," In: *15th Statistics Research Symposium Proceedings*, 2006, pp. 560-566.

[18]  A.W. Lo, "Long memory in stock market prices," *Econometrica,* vol. 59, pp. 1279-1313, 1991.

# CHAPTER 4

# Forecasting the Number of Outpatient Visits with Different Activation Functions

## Cagdas Hakan Aladag[1,*] and Sibel Aladag[2]

[1]*Hacettepe University, Faculty of Science, Department of Statistics, 06800, Ankara, Turkey and* [2]*General Director of Service Provision, Republic of Turkey Social Security Institution, Ankara, Turkey*

**Abstract:** Forecasting the number of outpatient visits plays important role in strategic decisions for the expert of healthcare administration. In order to manage hospitals effectively, it is needed to forecast the number of outpatient visits accurately. In the literature, there have been some methods proposed to forecast these time series. One of these methods is artificial neural networks approach. Although, artificial neural networks have proved its success in forecasting, there are still some problems with using this method. Determining the elements of this method is an important issue. Activation function is a crucial element of artificial neural networks. Therefore, in this study, we examined different activation functions to obtain more accurate out sample predictions while the number of patients is being forecasted. It is found that using different activation function affects the forecasting accuracy of feed forward neural network models.

**Keywords:** Activation function, Artificial neural networks, Forecasting, The number of outpatient visits, Time series.

## 1. INTRODUCTION

Forecasting activities play important role in our daily life. To make a strategic decision for the expert of healthcare administration, forecasting the number of outpatient visits plays important role. If more accurate forecasts are obtained, a schedule for human resources and finances can be made up more reasonably and hospital material resources can be distributed more suitably. In the literature, some approaches have been employed to obtain accurate forecasts for number of patients. Loytonen [1] used Box-Jenkins method to forecast HIV seropositive population in Finland. A stepwise autoregressive method and exponential smoothing models were used to forecast the number of patients with end-stage renal disease in the United States by Xue *et al.* [2]. A stepwise linear regression analysis was performed to predict patient visits to an urgent care clinic by Batal *et al.* [3]. Guan *et al.* [4] used artificial neural network (ANN) in forecasting the incidence of hepatitis A. Seasonal autoregressive integrated moving average, time series regression, exponential smoothing, and ANN were performed to forecast the daily patient volumes in the emergency department by Jones *et al.* [5]. Cheng *et al.* [6] used fuzzy time series to forecast number of outpatient visits.

Many time series in real life may have both linear and nonlinear structures. The traditional time series method can be inefficient to forecast such time series. ANN can model both linear and nonlinear structure in time series. For the aim of obtaining accurate forecasts in real life time series, ANNs have been used successfully in many applications in the literature. Feed forward neural network models have been especially preferred for forecasting time series in many implementations because of its easy usage and providing good results.

ANNs consist of a lot of elements such as architecture structure, training algorithm, activation function and there are various types of these elements [7]. When the number of patients is forecasted by employing ANNs, small amount of selections for elements of network models have been generally examined in the literature. Although the activation function has an important effect on the forecasting accuracy, in

---

*Address correspondence to Cagdas Hakan Aladag:* Hacettepe University, Faculty of Science, Department of Statistics, 06800, Ankara, Turkey; E-mail: chaladag@gmail.com

determining the best network model, different activation functions have not been tried in the analysis. Instead of employing different activation function, only one activation function has been usually utilized in the implementations. For example, Guan *et al.* [4] and Jones *et al.* [5] use only one type of activation function in their studies. Similarly, small amount of architecture is employed in the analysis. For example, Guan *et al.* [4] used 24 different architecture and Jones *et al.* [5] used only one structure including two neurons in hidden layer.

In this study, we use feed forward neural networks to forecast number of outpatient visits in health center at Hacettepe University. The monthly observations are between September 2004 and October 2008 so the time series include 50 observations. We examined different activation functions to obtain more accurate out sample predictions. Tangent sigmoid, logistic sigmoid, linear, satlin and triangular basis functions were used as activation functions in the output neuron while tangent sigmoid function is used in hidden layer neurons. We also examined 196 different architectures for each type of activation function. Also, the time series is analyzed by Box-Jenkins method, which is the well known method, for the comparison.

In the next section, brief information about ANNs is presented and some well known activation functions are given. Section 3 includes the implementation. As a result of the implementation, the obtained findings are summarized in the last section of the chapter.

## 2. ARTIFICIAL NEURAL NETWORKS

ANNs were originally motivated by the biological structures in the brains of humans and the animals, which are extremely powerful for such tasks as information processing, learning and adaptation. In forecasting, ANNs are mathematical models that imitate biological neural networks [8]. The function of an ANN is to produce an output pattern when presented with an input pattern. The elements of ANN are network architecture, learning algorithm and activation function.

The definition of ANN approach presented by Aladag *et al.* [9] can be given as follows:

'What is an artificial neural network?' is the first question that should be answered. Picton [10] answered this question by separating this question into two parts. The first part is why it is called an artificial neural network. It is called an artificial neural network because it is a network of interconnected elements. These elements were inspired from studies of biological nervous systems. In other words, ANN are an attempt at creating machines that work in a similar way to the human brain by building these machines using components that behave like biological neurons. The second part is what an artificial neural network does. The function of an artificial neural network is to produce an output pattern when presented with an input pattern. In forecasting, ANN are mathematical models that imitate biological neural networks. ANN consist of some elements. Determining the elements of the ANN issue that affect the forecasting performance of ANN should be considered carefully.

*Network architecture*: The structure of multilayer feed forward ANN is basically given in Fig. 1. Multilayer feed forward ANN as illustrated in the figure consist of three parts such as input, hidden and output layers. Each layer consists of neurons. The architecture structure is determined based on deciding the number of neuron in each layer. These neurons are linked each other by weights. There is no link among the neurons in the same layer [11].

*Learning algorithm*: There have been many learning algorithms in order to determine weights. The one of the most employed algorithm is called Back Propagation Learning Algorithm. This learning algorithm updates the weights based on difference between real value and output value of the ANN. However, back propagation networks have some disadvantages mentioned in the introduction. In light of the weakness of the conventional back propagation algorithm, a number of variations or modifications of this algorithm, such as the adaptive method, quickprop, and second-order methods *etc.*, have been proposed [12]. Among them, the second-order methods such as Levenberg Marquardt method are more efficient nonlinear optimization methods and are used in most optimization packages [13]. Their faster convergence, robustness, and the ability to find good local minima make them attractive in ANN training [12]. Therefore, Levenberg Marquardt method is used as training algorithm in the implementation.

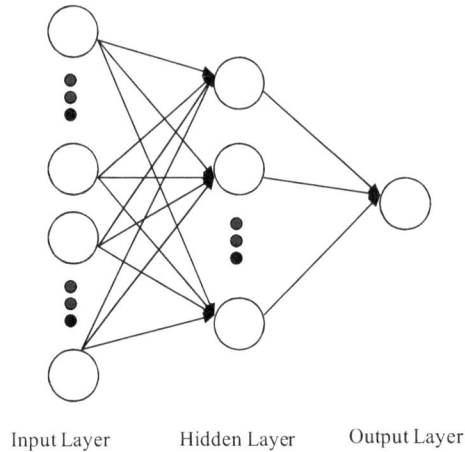

Input Layer          Hidden Layer          Output Layer

**Figure 1:** A broad feed forward ANN architecture.

*Activation function*: Activation function provides the non-linear mapping between input and output. The performance of networks depends on the proper choice of activation function. Activation function can be chosen as either linear or double polarized, or one polarized. Slope parameter should be determined when the activation is non-linear. Also, slope parameter plays a key role in reaching desired output values [11]. Some well known activation functions are tangent sigmoid, logistic sigmoid, linear, satlin, and triangular basis functions. In Matlab computer package, these functions are represented by tansig, logsig, purelin, satlin, and tribas, respectively. These activation functions, related formulas, and their graphs can be given as follows:

*Tangent sigmoid:*

$$a = tansig(n) = \frac{2}{1 + \exp(-2n)} - 1$$

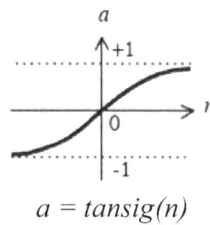

$$a = tansig(n)$$

*Logistic sigmoid:*

$$a = logsig(n) = \frac{1}{1 + \exp(-n)}$$

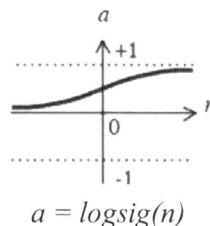

$$a = logsig(n)$$

*Linear:*

$$a = purelin(n) = n$$

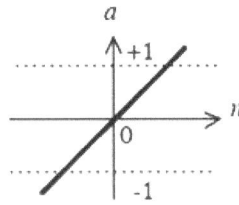

$$a = purelin(n)$$

*Satlin:*

$$a = satlin(n) = \begin{cases} 0 & n \leq 0 \\ n & 0 \leq n \leq 1 \\ 1 & 1 \leq n \end{cases}$$

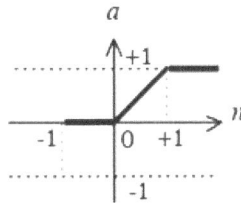

$$a = satlin(n)$$

*Triangular basis:*

$$a = tribas(n) = \begin{cases} 1 - |n| & -1 \leq n \leq 1 \\ 0 & otherwise \end{cases}$$

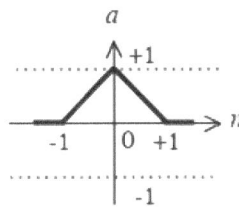

$$a = tribas(n)$$

In this study, we also employ these activation functions given above in the implementation. All of these activation functions are examined to get accurate forecasts for the number of outpatient visits time series.

## 3. THE IMPLEMENTATION

In the implementation, the number of outpatient visits in health center at Hacettepe University is forecasted by using feed forward neural networks. The monthly observations are between September 2004 and October 2008 so the time series include 50 observations. The graph of the time series can be seen in Fig. **2**.

Different network architectures are examined to obtain most accurate forecasts. Tangent sigmoid, logistic sigmoid, linear, satlin and triangular basis functions were used as activation functions in the output neuron while tangent sigmoid function is used in hidden layer neurons. We used feed forward neural network with input layer, one hidden layer and one output neuron in output layer. For each activation function, the number of neurons in input layer changes 1 through 14, the number of neurons in hidden layer alters 1 through 14 again, and one neuron exists in output layer so 196 architectures are used totally. The total number of 980 different architectures was examined for five different activation functions.

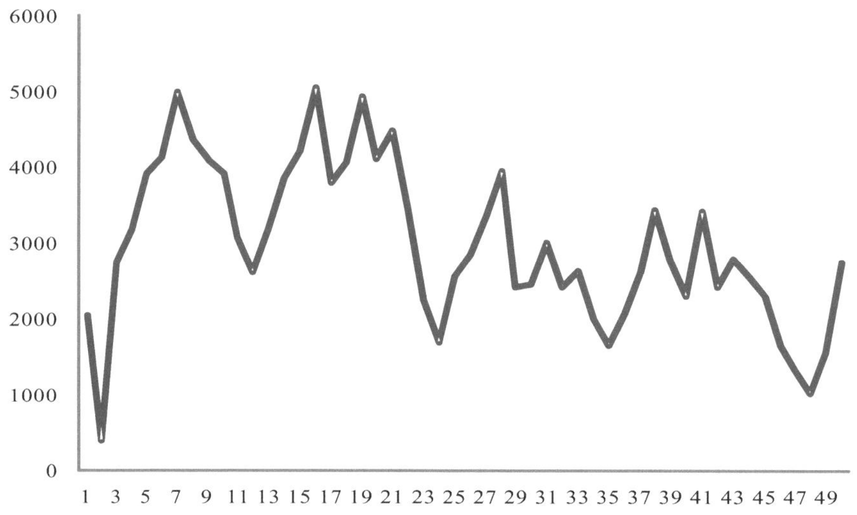

**Figure 2:** The graph of the time series.

Levenberg-Marquardt back propagation algorithm was used as learning algorithm since it is known as the fastest training algorithm. The first 45 and the last 5 observations were used for training and test sets, respectively. The best architectures were determined by using root of mean square error (RMSE) on test set. The RMSE criteria can be calculated as follows:

$$RMSE = \left(\frac{\sum_{i=1}^{T}(y_i - \hat{y}_i)^2}{T}\right)^{1/2}$$

where $y_i$ is the actual value, $\hat{y}_i$ is the predicted value and $T$ is the number of data.

**Table 1:** The best architectures

| Activation function | Best Arch. | RMSE |
|---|---|---|
| Tangent sigmoid | 3-5-1 | 203.06 |
| Logistic sigmoid | 13-6-1 | 243.28 |
| Linear | 6-2-1 | 222.36 |
| Satlin | 7-6-1 | 192.66 |
| **Triangular basis** | **8-2-1** | **135.83** |
| SARIMA(1,1,0)(0,1,0) | - | 401.58 |

When tangent sigmoid activation function was used, the architecture includes three input neurons, five neurons in hidden layer and one output neuron (3-5-1) was found as the best architecture with 203.06 RMSE value. We found the best architecture (13-6-1) with 243.28 RMSE value, (6-2-1) with 222.36 RMSE value, (7-6-1) with 192.66 and (8-2-1) with 135.83 RMSE value for logistic sigmoid, linear, satlin and triangular basis activation functions, respectively. It is observed from these results that triangular basis activation function gives the most accurate forecasts. When the time series was analyzed by Box-Jenkins method for comparison, SARIMA(1,1,0)(0,1,0) was found the best model with 401.58 RMSE values. It is clearly seen that all artificial neural network models give better results than Box-Jenkins method. Another important result is that using different activation function affects the forecasting accuracy of feed forward neural networks models. It is clearly seen in Table **1**, for the number of outpatient visits in health center at Hacettepe University, the most accurate forecasts are obtained by utilizing triangular basis activation. For the test set, the forecasts values obtained from the models given in Table **1** and the observed values are also summarized in Table **2**.

**Table 2:** The observed and obtained forecasts for test set

| Observed values | Tangent Sigmoid | Logistic Sigmoid | Linear | Satlin | Triangular basis | SARIMA |
|---|---|---|---|---|---|---|
| 1651 | 1910,77 | 1785,35 | 1453,88 | 1588,57 | 1748,78 | 1795,77 |
| 1330 | 1374,67 | 1534,86 | 1626,10 | 1395,70 | 1465,95 | 1271,84 |
| 1020 | 1316,15 | 1337,73 | 1218,35 | 1106,98 | 948,44 | 1684,07 |
| 1568 | 1379,61 | 2068,87 | 1803,31 | 1886,58 | 1785,60 | 2131,49 |
| 2746 | 2629,81 | 2545,43 | 2907,15 | 3007,39 | 2637,61 | 2899,21 |

In order to show the results visually the real values and the forecasts obtained from the SARIMA model and the constructed ANNs models presented in Table **1** are shown in Figs. **3**–**7**.

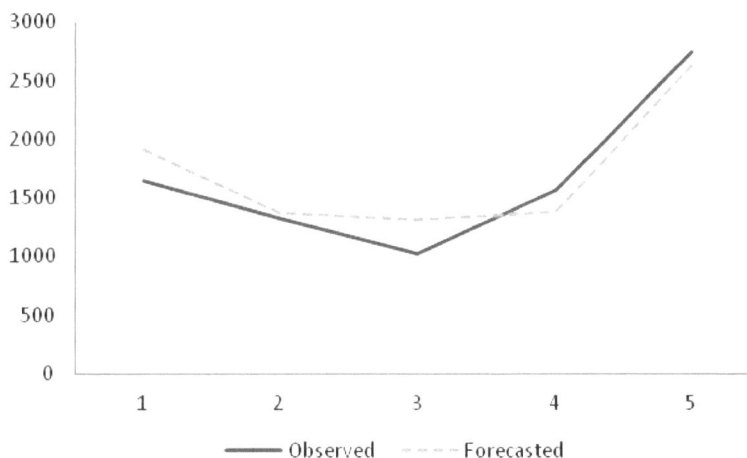

**Figure 3:** The graph of the real and the forecast values obtained from tangent sigmoid function.

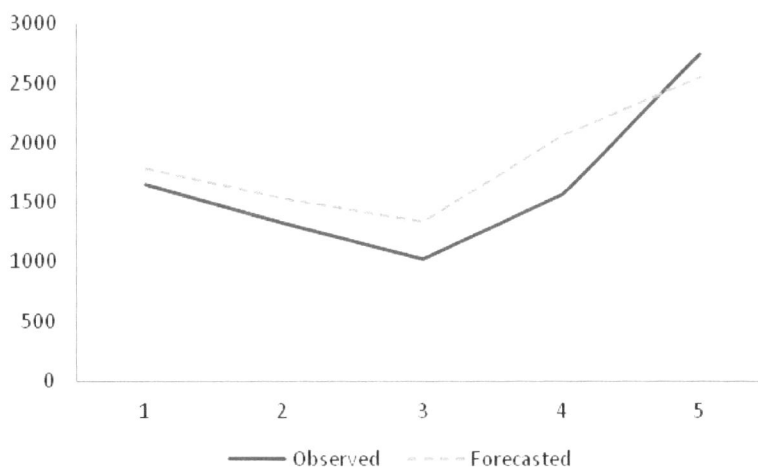

**Figure 4:** The graph of the real and the forecast values obtained from logistic sigmoid function.

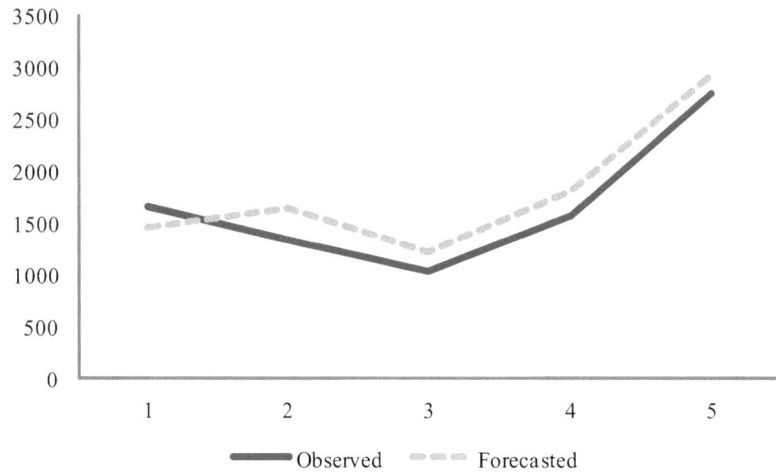

**Figure 5:** The graph of the real and the forecast values obtained from linear function.

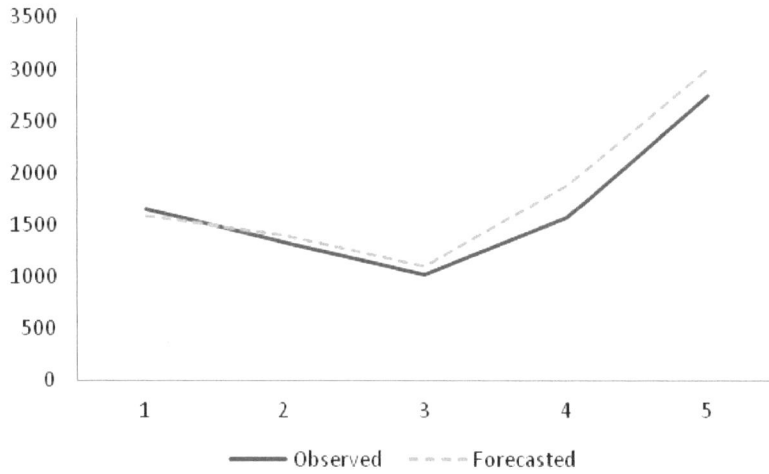

**Figure 6:** The graph of the real and the forecast values obtained from satlin function.

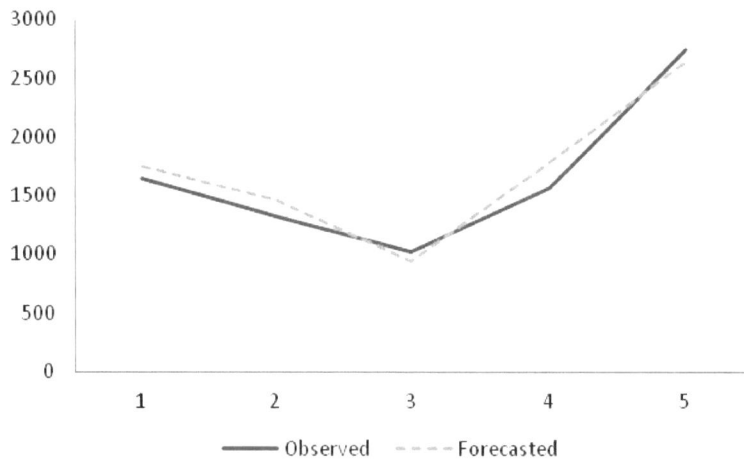

**Figure 7:** The graph of the real and the forecast values obtained from triangular basis function.

## 4. CONCLUSION

The traditional time series method can be inefficient to forecast by reason of having both linear and nonlinear structures. ANN can model both linear and nonlinear structure in time series. To obtain accurate forecasts, ANNs have been used successfully in many applications in the literature.

ANNs consist of a lot of elements such as architecture structure, training algorithm, activation function and there are various types of these elements. It is important to determine the best architecture which gives the most accurate forecasts when ANN are used to forecast time series. Activation function provides the non-linear mapping between input and output. The performance of networks depends on the proper choice of activation function. In order to determine the best architecture, activation function plays an important role. Using different activation function affects the forecasting accuracy of feed forward neural networks models.

In this study, the number of outpatient visits in health center at Hacettepe University is forecasted by using feed forward neural networks. Different network architectures are examined to obtain most accurate forecasts. Tangent sigmoid, logistic sigmoid, linear, satlin and triangular basis functions were used as activation functions in the output neuron while tangent sigmoid function is used in hidden layer neurons. All artificial neural network models give better results than Box-Jenkins method. Another important result is that using different activation function affects the forecasting accuracy of feed forward neural networks models.

## REFERENCES

[1]     M. Loytonen, "The Box-Jenkins forecast of HIV seropositive population in Finland, 1991-1993," *Geografiska Annaler*, vol. 73, no. 2, pp. 121-131, 1991.

[2]     J.L. Xue, J.Z. Ma, T.A. Louis, and A.J. Collins, "Forecast of the number of patients with end-stage renal disease in the United States to the year 2010," *J Am Soc Nephrol*, vol. 12, pp. 2753–2758, 2001.

[3]     H. Batal, J. Tench, S. McMillan, J. Adams, and P.S. Mehler, "Predicting patient visits to an urgent care clinic using calendar variables," *Acad Emerg Med*, vol. 8, no. 1, pp. 48-53, 2001.

[4]     P. Guan, D.S. Huang, and B.S. Zhou, "Forecasting model for the incidence of hepatitis A based on artificial neural network," *World J Gastroentero*, vol. 10 (24), pp. 3579-3582, 2004.

[5]     S.S. Jones, A. Thomas, S. Evans, S.J. Welch, P.J. Haug, and G.L. Snow, "Forecasting daily patient volumes in the emergency department," *Acad Emerg Med*, vol. 15, pp. 159-170, 2008.

[6]     C.H. Cheng, J.W. Wang, and C.H. Li,, "Forecasting the number of outpatient visits using a new fuzzy time series based on weighted-transitional matrix," *Expert Syst Appl*, vol. 34, pp. 2568-2575, 2008.

[7]     E. Eğrioglu, C.H. Aladag, and S. Gunay, "A new model selection strategy in artificial neural network" *Appl Math Comput*, vol. 195, pp. 591-597, 2008.

[8]     E. Eğrioglu, C.H. Aladag, U. Yolcu, V.R. Uslu, M.A. Basaran, "A new approach based on artificial neural networks for high order multivariate fuzzy time series," *Expert Syst Appl*, vol. 36, no. 7, pp. 10589-10594, 2009.

[9]     C.H. Aladag, E. Eğrioglu, and S. Gunay, "A new architecture selection strategy in solving seasonal autoregressive time series by artificial neural networks," *Hacet J Math Stat*, vol. 37, no. 2, pp. 185-200, 2008.

[10]   P.D. Picton, *Introduction to Neural Networks*, Macmillan Press Ltd., 1994.

[11]   E. Eğrioglu, C.H. Aladag, V.R. Uslu, M.A. Basaran, and U. Yolcu, "A new hybrid approach based on SARIMA and partial high order bivariate fuzzy time series forecasting model," *Expert Syst Appl*, vol. 36, no. 4, pp. 7424-7434, 2009.

[12]   H.F. Zou, G.P. Xia, F.T. Yang, and H.Y. Wang, "An investigation and comparison of artificial neural network and time series models for Chinese food grain price forecasting," *Neurocomputing*, vol. 70, pp. 2913-2923, 2007.

[13]   C.H. Aladag, "A new architecture selection method based on tabu search for artificial neural networks, *Expert Syst Appl*, vol. 38, pp. 3287-3293, 2011.

*Advances in Time Series Forecasting*, 2012, 34-39

# CHAPTER 5

## Adaptive Weighted Information Criterion to Determine the Best Architecture

## Cagdas Hakan Aladag[1,*] and Erol Eğrioğlu[2]

*[1]Hacettepe University, Faculty of Science, Department of Statistics, 06800, Ankara, Turkey and [2]Ondokuz Mayis University, Faculty of Arts and Science, Department of Statistics, 55139, Samsun, Turkey*

**Abstract:** In the literature, different selection criteria are used for determining the best architecture when time series is analyzed by artificial neural networks. Criteria available in the literature measure different properties of forecasts. To obtain better forecasts, Eğrioğlu *et al.* [1] proposed a criterion which can measure all properties of forecasts. Aladag *et al.* [2] improved the criterion proposed by [1] by using optimization. In this study, both the weighted information criterion proposed by Eğrioğlu *et al.* [1] and the adaptive weighted information criterion proposed by Aladag *et al.* [2] are introduced. These criteria are used in the architecture selection to analyze time series which are the import values of Turkey and the air pollution records in Ankara. As a result of computations, obtained results are compared and discussed. As a result of the comparison, it is seen that adaptive weighted information criterion produce more consistent results.

**Keywords:** Artificial neural networks, Forecasting, Model selection criterion, Time series.

## 1. INTRODUCTION

Artificial neural networks (ANN) offer effective and alternative solutions to the problems which are the subjects of statistics. Recently, there have been numerous studies using ANN approaches for time series forecasting problem. It is well known that ANN method is effective particularly in modeling non-linear time series. Not requiring any assumption in forecasting time series makes ANN applicable for many fields. Although ANN has significant advantages, there are still problems to be solved. These problems can be expressed as follows:

- Which architecture structure should be used?

- Which activation function should be used?

- How many hidden layers should be used?

- How many units should hidden and input layers should have?

- Which algorithm should be preferred for ANN training?

- Which model selection criterion should be used for the evaluation of the forecasts?

- Which ANN type (Elman, Jordan, multi-layer) should be used?

There have been partly answers to these questions in the literature. In fact, these questions are valid not only for forecasting but also other targets such as classification. In the literature, [3] partly cover answers to these questions. It is possible to reach following conclusions from the studies available in the literature using artificial neural networks for time series forecasting [4]:

---

*Address correspondence to Cagdas Hakan Aladag: Hacettepe University, Faculty of Science, Department of Statistics, 06800, Ankara, Turkey; E-mail: chaladag@gmail.com

- Multi-layer ANN gives successful results for forecasting problem.

- It is possible to create accurate forecasting with the use of single hidden layer.

- For satisfactory results, logistic or hyperbolic tangent activation function can be used as an activation function in all layers.

- Linear activation function can also be preferred in output layer.

Despite all these experimental results, the problem which lacks of a common solution and a systematic approach is the selection of the number of units to be used in input and hidden layers. As hidden layer units show delayed-variables in time series forecasting problem, determination of the number of units in input layer is a factor affecting forecasting performance directly. If the number of units in hidden layer is assessed accurately, modeling ability of the network will be optimized. Therefore, determination of hidden layer and input layer unit numbers is crucial. On the other hand, in the literature it is still problematic to assess the best unit number. This is called architecture style selection problem. In the literature, systematic methods not ensuring optimal solution have been suggested for architecture selection problem. Lahnajarvi *et al.* [5], Reed [6] and Siestema and Dow [7] proposed various pruning algorithms. Buhamra *et al.* [8] suggested an input layer unit number selection method based on Box-Jenkins approach. Aladag [9] and Dam and Saraf [10] proposed architecture selection methods based on tabu search algorithm and genetic algorithms, respectively. Aladag *et al.* [11] developed a new approach for architecture selection in analyzing seasonal time series with ANN methods. Zeng *et al.* [12] proposed an approach based on principal component analysis. Zhang [3] developed iterative construction algorithm. The approaches proposed in the studies [13-18] are the examples of other systematic approaches in the literature.

Eğrioğlu *et al.* [1] proposed weighted information criterion assessing forecasting in terms of growth of the error, the number of parameters and direction accuracy in the model. Roughly, weighted information criterion (WIC) is obtained by summing some weighted different selection criteria which measure the forecasting accuracy of an ANN model in different ways. Aladag *et al.* [2] improved this criterion by determining the weights used for WIC with the use of optimization. In this study, both weighted information criterion proposed by Eğrioğlu *et al.* [1] and adaptive weighted information criterion proposed by Aladag *et al.* [2] were introduced. In this application, the best ANN architecture was determined using both WIC and adaptive weighted information criterion (AWIC) in time series analysis of import values of Turkey and the air pollution records in Ankara. Finally results obtained from the application were compared and discussed.

## 2. WEIGHTED INFORMATION CRITERION

WIC proposed by Eğrioğlu *et al.* [1], is the weighted sum up of most preferred criteria including Akaike information criterion (AIC), Bayesian information criterion (BIC), root mean squared error (RMSE), mean absolute percentage error (MAPE), directional accuracy (DA) and modified direction accuracy. Some of the criteria are calculated by the use of following formulas.

$$AIC = \log\left(\frac{\sum_{i=1}^{T}(y_i - \hat{y}_i)^2}{T}\right) + \frac{2m}{T} \qquad (1)$$

$$BIC = \log\left(\frac{\sum_{i=1}^{T}(y_i - \hat{y}_i)^2}{T}\right) + \frac{m\log(T)}{T} \qquad (2)$$

$$RMSE = \left( \frac{\sum_{i=1}^{T}(y_i - \hat{y}_i)^2}{T} \right)^{1/2} \tag{3}$$

$$MAPE = \frac{1}{T} \sum_{i=1}^{T} \left| \frac{y_i - \hat{y}_i}{y_i} \right| \tag{4}$$

$$DA = \frac{1}{T} \sum_{i=1}^{T} a_i \quad , \quad a_i = \begin{cases} 1 \quad , & (y_{i+1} - y_i)(\hat{y}_{i+1} - y_i) > 0 \\ 0 \quad , & otherwise. \end{cases} \tag{5}$$

In the formulas, $y_i$ represents real values; $\hat{y}_i$ forecasts obtained from ANN architecture, $T$, observation number in test set and $m$, weight number in ANN architecture. Another criterion, called modified direction accuracy (MDA) assessing that how well the milestones were forecasted, was proposed by Eğrioğlu *et al.* [1]. ADS criterion can be calculated as;

$$A_i = 1 \quad , \quad y_{i+1} - y_i \leq 0$$
$$A_i = 0 \quad , \quad y_{i+1} - y_i > 0$$

$$F_i = 1 \quad , \quad \hat{y}_{i+1} - \hat{y}_i \leq 0$$
$$F_i = 0 \quad , \quad \hat{y}_{i+1} - \hat{y}_i > 0$$

$$D_i = (A_i - F_i)^2$$

$$MDA = \frac{\sum_{i=1}^{T-1} D_i}{T-1} \tag{6}$$

Accordingly, architecture selection strategy based on WIC is presented in 5 steps in the following algorithm introduced in [1].

***Step 1.*** Possible architectures are defined. For example, when both input layer unit number and hidden layer unit number vary 1 and 12, it is possible to create a total of 144 architectures.

***Step 2.*** The best values of ANN are determined using training data and AIC, BIC, RMSE, MAPE, DA and MDA criteria were calculated for the test data.

***Step 3.*** AIC, BIC, RMSE, MAPE, DA and MDA criteria calculated for all possible architectures are standardized. For example, AIC criterion for 144 architectures is standardized as follows;

$$AIC_i = \frac{AIC_i - \min(AIC)}{\max(AIC) - \min(AIC)}, \quad i = 1,...,144$$

***Step 4.*** Weighted information criterion (WIC) is calculated with the following formula for each architecture:

$$WIC = 0,1(AIC + BIC) + 0,2(RMSE + MAPE) + 0,2((1 - DA) + MDA) \tag{7}$$

***Step 5.*** Architecture with minimum WIC value is selected.

Eğrioğlu *et al.* [1] heuristically determined the weights in the equation (6). As AIC and BIC criteria choose models with minimum parameters, WIC weights were taken as 0.1. Weights of other criteria were selected coequally as 0.2.

## 3. ADAPTIVE WEIGHTED INFORMATION CRITERION (AWIC)

In their study, Aladag *et al.* [2] improved WIC which was proposed by Eğrioğlu *et al.* [1] by determining the criterion weights with the use of optimization method. Calculated new criterion based on WIC was named AWIC. AWIC is calculated determining the WIC weights that maximize the correlation between test sets. Aladag *et al.* (2010) showed that AWIC criterion was more consistent than that in WIC. In order to calculate the AWIC criterion, analyzed time series are divided in two as training and test sets and test set is divided in two sub-groups. For both test sets, AIC, BIC, RMSE, MAPE, DA and MDA criteria are calculated for different architectures. Then, these criteria are standardized in a way to be between (0, 1). $AWIC_1$ and $AWIC_2$ criteria for the first and second test sets are calculated by the following formulas;

$$AWIC_1(i) = w_1 RMSE_1(i) + w_2 MAPE_1(i) + w_3(1 - DA_1(i)) + w_4 MDA_1(i)$$
$$+ 0,1 AIC_1(i) + 0,1 BIC_1(i), \ i = 1, 2, ..., m_i \times m_h$$

$$AWIC_2(i) = w_1 RMSE_2(i) + w_2 RMSE_2(i) + w_3(1 - DA_2(i)) + w_4 MDA_2(i)$$
$$+ 0,1 AIC_2(i) + 0,1 BIC_2(i), i = 1, 2, ..., m_i \times m_h$$

Here, $w_j$ ($j=1,2,3,4$) represents weights and coefficients, $m_i$ and $m_h$ represent input layer and hidden layer unit number respectively. As output layer has only one unit, $m_i$ and $m_h$ multiplication gives the total architecture number. For example, $AWIC_1(i)$ is the calculated AWIC value based on the first set for $i^{th}$ architecture. $AWIC_2(i)$ is the calculated AWIC criterion value based on the second set for $i^{th}$ architecture.

Due to their high consistency, the coefficients of AIC and BIC criteria were excluded from the optimization and their weights are fixed with the value of 0,1 [2]. During the optimization process, weights are optimized to be closest to correlation coefficient +1 between $AWIc_1(i)$ and $AWIc_2(i)$. If the calculated correlation coefficient is close to +1 value, the more consistent value for two sets of AWIC is obtained. In another words, more consistent architecture selection criterion means better forecast results [2].

## 4. APPLICATION

In the application, import values of Turkey (IVT) between the years 1923 and 2003, and the air pollution records in Ankara between January 1996 and October 2002 (ANSO) were used. Annual import data and air pollution records consist of 81 and 95 observation respectively. Fifteen percent of the time series were taken as test set and then test set was divided in to three equal sets. AWIC weights were determined on first two test sets by optimization. Then, results obtained from WIC and AWIC were compared on the third test set. Possible architecture number is defined as 144 by varying input unit and hidden layer unit numbers between 1 and 12. The correlation between values received from the first and second test sets for analyzed 144 architectures was $r_1$ and AWIC coefficients were optimized to maximize this correlation. The correlation between values received from the second and third test sets was $r_2$ and used to compare AWIC and WIC consistency. In short, $r_2$ represents the correlation between $AWIc_2(i)$ and $AWIc_3(i)$ values. The formula used for the calculation of $AWIc_3(i)$ value is given below;

$$AWIC_3(i) = w_1 RMSE_3(i) + w_2 MAPE_3(i) + w_3(1 - DA_3(i)) + w_4 MDA_3(i)$$
$$+ 0.1 AIC_3(i) + 0.1 BIC_3(i), i = 1, 2, ..., m_i \times m_h$$

Results obtained from calculation of import values and Ankara air pollution records are summarized in Tables **1** and **2**. Table **1** covers the best architectures and correlation coefficients obtained from WIC and AWIC. Table **2** shows the AWIC weight values obtained at the end of the optimization. When calculated $r_2$

correlation coefficients were analyzed, as it is clearly seen in Table **1**, AWIC gives more consistent results for both time series.

**Table 1:** Results obtained from WIC and AWIC

| Criterion | DATA | $r_1$ | $r_2$ | The Best Architecture |
|---|---|---|---|---|
| WIC | IVT | 0.6515 | 0.7620 | 2-3-1 |
| | ANSO | 0.5455 | 0.5951 | 6-2-1 |
| AWIC | IVT | 0.7507 | 0.8922 | 2-2-1 |
| | ANSO | 0.7112 | 0.7112 | 4-5-1 |

**Table 2:** The optimal weights of calculated AWIC

| Data | $w_1$ | $w_2$ | $w_3$ | $w_4$ | $w_5$ | $w_6$ |
|---|---|---|---|---|---|---|
| IVT | 0.8000 | 0.0000 | 0.0000 | 0.0000 | 0.1000 | 0.1000 |
| ANSO | 0.7603 | 0.0000 | 0.0000 | 0.0397 | 0.1000 | 0.1000 |

## 5. RESULTS

In this study, the use of both the weighted information criterion proposed by Eğrioğlu *et al.* [1] and the adaptive weighted information criterion proposed by Aladag *et al.* [2] in the determination of the best ANN architecture was explained. In the implementation section, import values of Turkey and air pollution records in Ankara time series were analyzed and obtained results were compared. When the correlation coefficients calculated from the times series were analyzed, it was seen that consistency of AWIC is greater than that of WIC for these real time series.

## REFERENCES

[1]   E. Eğrioğlu, C.H. Aladag, and S. Gunay, "A new model selection strategy in artificial neural networks," *Appl Math Comput*, vol. 195, pp. 591-597, 2008.

[2]   C.H. Aladag, E. Eğrioğlu, and S. Gunay, M.A. Basaran, "Improving weighted information criterion by using optimization," *J Comput Appl Math*, vol. 233, pp. 2683-2687, 2010.

[3]   G.P. Zhang, B.E. Patuwo, and Y.M. Hu, Forecasting with Artificial Neural Networks: The State of the Art. *Int J Forecasting*, vol. 14, pp. 35-62, 1998.

[4]   S. Gunay, E. Eğrioğlu, and C.H. Aladag, *Introduction to Single variable Time Series Analysis*. Hacettepe University Press: Ankara, 2007.

[5]   J.J.T. Lahnajarvi, M.I. Lehtokangas, and J.P.P. Saarinen, "Evaluation of constructive neural networks with cascaded Architectures," *Neurocomputing*, vol. 48, pp. 573-607, 2002.

[6]   R. Reed, "Pruning algorithms a survey," *IEEE Trans Neural Netw*, vol. 4, pp. 740-747, 1993.

[7]   J. Siestema, and R. Dow, "Neural net pruning – why and how?" *Proceedings of the IEEE International Conference on Neural Networks*, 1988, pp. 325-333.

[8]   S. Buhamra, N. Smaoui, and M. Gabr, "The Box–Jenkins analysis and neural networks: and time series modeling," *Appl Math Model*, vol. 27, pp. 805-815, 2003.

[9]   C.H. Aladag, "Using tabu search algorithm in the selection of architecture for artificial neural networks", Ph.D. thesis, Hacettepe University, Turkey, 2009.

[10]  M. Dam, and D.N. Saraf, "Design of neural networks using genetic algorithm for on-line property estimation of crude fractionator products," *Comput Chem Eng*, vol. 30, no. 4, pp. 722-729, 2006.

[11]  C.H. Aladag, E. Eğrioğlu,and S. Gunay, "A new architecture selection strategy in solving seasonal autoregressive time series by artificial neural networks," *Hacet J Math Stat*, vol. 37, no. 2, pp. 185-200, 2008.

[12]  J. Zeng, H. Guo, and Y. Hu, "Artificial neural network model for identifying taxi gross emitter from remote sensing data of vehicle emission," *J Environ Sci*, vol. 19, pp. 427-431, 2007.

[13]  T.F. Rathbun, S.K. Rogers, M.P. DeSimio, and M.E. Oxley, "MLP iterative construction algorithm," *Neurocomputing*, vol. 17, no. 4, pp. 195-216, 1997.

[14]   B. Durbin, S. Dudoit, and M.J. Van Der Laan, "A deletion/substitution/addition algorithm for classification neural networks, with applications to biomedical data," *J Stat Plan Infer*, vol. 138, pp. 464-488, 2008.

[15]   H.C. Yuan, F.L. Xiong, and X.Y. Huai, "A method for estimating the number of hidden neurons in feed-forward neural networks based on information entropy," *Comput Electron Agr*, vol. 40, pp. 57-64, 2003.

[16]   Z. Wang, C.D. Massimo, M.T. Tham, and A.J. Morris, "A procedure for determining the topology of multilayer feedforward neural networks," *Neural Networks*, vol. 7, pp. 291-300, 1994.

[17]   N. Murata, S. Yoshizawa, and S. Amari, "Network information criterion- determining the number of hidden units for an artificial neural network model," *IEEE Trans Neural Netw*, vol. 5, pp. 865-872, 1994.

[18]   A. Roy, L.S. Kim, and S. Mukhopadhyay, "A polynomial time algorithm for the construction and training of a class of multilayer perceptrons," *Neural Networks*, vol. 6, pp. 535-545, 1993.

# CHAPTER 6

# Public Expenditure Forecast by Using Feed Forward Neural Networks

## Alparslan A. Basaran[1], Cagdas Hakan Aladag[2,*], Necmiddin Bagdadioglu[1] and Suleyman Gunay[2]

*[1]Hacettepe University, Faculty of Economics and Administrative Sciences, Department of Public Finance, 06800, Ankara, Turkey and [2]Hacettepe University, Faculty of Science, Department of Statistics, 06800, Ankara, Turkey*

**Abstract:** The accurate forecast of public expenditure is crucial for the success of the new public financial management approach developed in Turkey since the financial crisis of 2001. The public institutions are now obliged to align their expenditure with the framework shaped by the Public Financial Management and Control Law (No: 5018), the Middle-Term Programme of 2010-2012, and recently the Fiscal Rule envisaged to apply in the next budgetary period. This necessitates a better forecasting method than the traditional way of budget forecasting, which is typically based on the expenditures of previous years adjusted by inflation. Particularly focusing on the expenditure side of the budget, this chapter applies various artificial neural networks models to the expenditures of 1973-2008 of two Turkish public institutions, namely, the State Planning Organization and the Court of Accounts to achieve accurate forecast levels. The artificial neural networks approach is rarely applied for the forecasting of public expenditures, and as far as we know this is the first of such attempts involving Turkish data. The artificial neural networks application provided very accurate public expenditure forecasts for these public institutions, suggesting that the artificial neural networks is a very useful method for the public expenditure forecasting, as well.

**Keywords:** Artificial neural networks, Budget forecasting, Public expenditure, Time series.

## 1. INTRODUCTION

In recent years, the artificial neural networks approach has been applied to many areas; one of them is the time series forecasting [1]. Since the artificial neural networks can model both nonlinear and linear structure of time series, using the artificial neural networks in forecasting can give more accurate results than the other methods [2]. The accurate forecast of public expenditure has crucial importance for Turkey considering her chronic budget imbalances derived from excess expenditures over collected revenues. A recent study by Bagdigen [3] suggests that the severe errors in budget forecasting in Turkey are caused by the under-forecasting of expenditures and the over-forecasting of revenues.

Turkey has shown her willingness for not tolerating such budget imbalances anymore by initiating an economic recovery programme, including a new public financial management approach since the financial crisis of 2001 [4]. As part of the economic recovery programme, presently, the public institutions are obliged to align their expenditure with the framework shaped by the Public Financial Management and Control Law (No: 5018), the Middle-Term Programme of 2010-2012, and recently the Fiscal Rule envisaged to apply in the next budgetary period. The success of this initiative necessitates a better forecasting method than the traditional method of budget forecasting in Turkey, which is typically based on the adjustment of expenditures and revenues of previous years by inflation.

The aim of this chapter is to show how useful the artificial neural networks can be in getting more accurate public budget expenditure forecasts in Turkey. The artificial neural networks approach is rarely applied for the forecast of public expenditures [5-7], and as far as we know this is the first of such attempts involving Turkish data. The chapter is organized under three main sections. The following section

*\*Address correspondence to Cagdas Hakan Aladag:* Hacettepe University, Faculty of Science, Department of Statistics, 06800, Ankara, Turkey; E-mail: chaladag@gmail.com

introduces the artificial neural networks. The third section explains the basics of the new budgetary approach in Turkey, and the involvement of the State Planning Organization (SPO) and the Court of Accounts (CoA) in the budgetary process. The fourth section applies the artificial neural networks on the expenditures of the SPO and the CoA. The chapter ends with an assessment of the findings.

## 2. THE ARTIFICIAL NEURAL NETWORKS

Aladag *et al.* [8] gave brief information about the artificial neural networks as follows:'What is an artificial neural network?' is the first question that should be answered. Picton [9] answered this question by separating this question into two parts. The first part is why it is called as artificial neural network. It is called as artificial neural network because it is a network of interconnected elements. These elements were inspired from studies of biological nervous systems. In other words, the artificial neural networks are an attempt at creating machines that work in a similar way to the human brain by building these machines using components that behave like biological neurons.

The second question is what an artificial neural network does? The function of an artificial neural network is to produce an output pattern when presented with an input pattern. In forecasting, the artificial neural networks are mathematical models that imitate the biological neural networks. The artificial neural networks consist of some elements. Determining the elements of the artificial neural networks issue that affect the forecasting performance of artificial neural networks should be considered carefully. Elements of the artificial neural networks are generally given as network architecture, learning algorithm and activation function [10].

One critical decision is to determine the appropriate architecture, that is, the number of layers, the number of nodes in each layers and the number of arcs which interconnects with the nodes [11]. However, in the literature, there are not general rules for determining the best architecture. Therefore, several architectures should be tried for the correct results. There are various types of artificial neural networks. One of them is called as feed forward neural networks. The feed forward neural networks have been used successfully in many studies [10]. In the feed forward neural networks, there are no feedback connections. The broad feed forward neural network architecture that has single hidden layer and single output is given as an illustration in Fig. **1**.

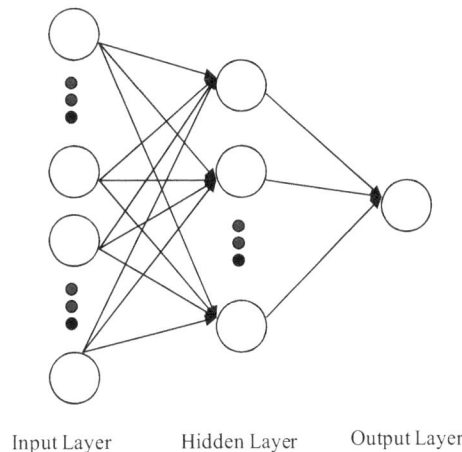

Input Layer     Hidden Layer     Output Layer

**Figure 1:** A broad feed forward neural network architecture.

Learning of an artificial neural network for a specific task is equivalent to finding the values of all weights such that the desired output is generated by the corresponding input. Various training algorithms have been used for the determination of the optimal weights values. The most popularly used training method is the back propagation algorithm presented by Smith [12]. In the back propagation algorithm, learning of the artificial neural networks consists of adjusting all weights considering the error measure between the desired output and actual output [13].

Another element of the artificial neural networks is the activation function. It determines the relationship between inputs and outputs of a network. In general, the activation function introduces a degree of the non-linearity that is valuable in most of the artificial neural networks applications. The well known activation functions are logistic, hyperbolic tangent, sine (or cosine) and the linear functions. Among them, logistic activation function is the most popular one [14].

The artificial neural networks approach, as indicated above, has been rarely used for public expenditure forecasting particularly in Turkey. Before the application of the artificial neural networks on public expenditures forecasting, the next section briefs about the new budgetary approach recently become the norm by the publications of the Public Financial Management and Control Law (No: 5018) in Turkey.

## 3. THE NEW BUDGETARY APPROACH IN TURKEY

By means of the Public Financial Management and Control Law (No: 5018) published in 2004 and the switch on the Middle-Term Performance Based Budgetary approach, Turkey has taken important steps to attain the budgetary and financial discipline in the public sector. Undoubtedly the most important of these steps is the initiation of performance based budgetary process designed according to the strategic plan of each public institution. The strategic plan is prepared for five years in the light of the public institution's mission, vision, medium and long term goals. The strategic plan includes performance programme, performance indicators and monitoring systems which will be used to assess the achievements of the related public institution. The performance programme is the yearly reflection of the strategic plan, assessing whether the public resources are used in effective, economic and efficient way during the year. The strategic plan is prepared following a participatory process of shareholders and regarded as the main document establishing the link between the future perspective, performance and budget allowances of the public institution.

In the performance based budgeting approach, the central government guides the public institutions by two main documents used for the determination of macroeconomic goals and the budget allowances limitations. These are the Middle-Term Programme and the Middle-Term Plan prepared by the SPO and the Ministry of Finance (MoF), respectively. The Middle-Term Programme focuses on the macroeconomic indicators, while the Middle-Term Plan defines the limits of allowances that the public institutions ought to take into consideration.

Under the performance based budgeting approach there are three main documents determining the boundary of expenditures of the public institutions. These are the middle-term programme, the middle-term financial plan, and the strategic plan and performance programme. According to the Public Financial Management and Control Law (No: 5018), the middle-term programme prepared by the SPO determines the basic economic thresholds by taking into consideration of general economic conditions, strategic and development plans. Similarly, the middle-term plan prepared in line with the middle-term programme by the MoF defines the budget revenues and expenditures, the likely amounts of deficit/surplus, and the budget allowances ceilings for the next three years.

Ultimately, the public institutions are expected to forecast their expenditures as performance based and in line with the five year strategic plans and the yearly performance programmes. Meanwhile, these expenditure forecasts are expected to be compatible with the middle-term programme and the middle-term financial plan, as well. Unlike the classical budgetary approach, the performance based budgetary approach is pursued in a participatory involvement of shareholders in the budget process.

The accuracy of budget forecasts is also important in terms of two budgeting principles, namely, accuracy and sincerity, concerning the main stages of budget preparation, implementation and control. The principle of accuracy requires that the budget is prepared accordingly to the current macroeconomic situation. That is the budget ought to include objective forecasts of both revenues and expenditures. The principle of sincerity on the other hand requires that those who prepare the budget should not determine the revenues and the expenditures more or less than needed.

Through the Public Financial Management and Control Law (No: 5018), the budget forecasts are expected to be more consistent and predictable so that budgeting for multiple years can be performed. In this framework, the budgets of two important institutions of the central government, namely, the SPO and the CoA are analyzed. The choice of these institutions is not arbitrary. These institutions are chosen since they have important roles during the preparation and control of budget, and compared with the other public institutions they have rarely exceeded their allowed expenditures [15] as shown in the next subsection.

Besides the preparation of the Middle-Term Programme, the Public Financial Management and Control Law (No: 5018) assigns the SPO with various other duties. Firstly, the SPO is to determine the timetable for the process of strategic planning, and the procedure and guidelines for the strategic plans. Secondly, although the SPO has not had any participation within the process of performance based budgeting yet, both the SPO and the MoF are expected to work together in the future during both the determination of the performance indicators and their control. Thirdly, during the budget preparation process, the SPO is responsible from the publication of the Investment Circular and its addendum the Investment Programme Preparation Guide in the Official Gazette by the end of June of that year. The SPO prepares the Public Investment Programme of the Central Government in line with the Budget Law, and it collaborates with the MoF in the determination process of project allowances concerning public investments. Moreover, the SPO is the approval authority of the feasibility of public projects, and is also determining the procedure and principles concerning the implementation and monitoring of the investments of social security institutions and local authorities. Last but not least, the results of implementations and realizations of the public investment projects are examined by the SPO together with the MoF. In a nutshell, the SPO is one of the most important and influential public institutions regarding the determination of the level of public investments and the middle and long term targets of budgetary process in Turkey.

On the other hand, the CoA is performing a very important role in the judiciary and executive control of the public budget. The CoA is responsible for both the preparation of the General Conformity Statements for public institutions affiliated with the central government and their external control as stated in the Public Financial Management and Control Law (No: 5018):'Article 68- The purpose of the ex post external audit to be performed by the Court of Accounts is to audit, within the framework of the accountability of public administrations within the scope of general government, the financial activities, decisions and transactions of management in terms of their compliance with the laws, institutional purposes, targets and plans, and to report their results to the Turkish Grand National Assembly.

By this new budgetary approach the public institutions are persuaded to use their budget allowances in effective, economic and efficient way in the middle-term (three years). This brings with it the need for accurate budget forecasting and predictability of budget allowances in the middle-term. This is urgent for a country such as Turkey characterized by a bad record of chronic budget imbalances derived from excess expenditures over collected revenues. In this chapter the focus is on the expenditure side of the budget. However, the forecast of public revenues is by no means less important for the Turkish case. Our motivation to focus on the expenditures side of the budget is fuelled by the fact that Turkey has been introduced more than 20 tax reforms so far to adjust the revenue side of her public budget, while perhaps due to some political necessities neglected the expenditure side until the announcement of Public Financial Management and Control Law (No: 5018). The next section presents how the artificial neural networks approach can be used by the Turkish public institutions to determine their expenditures in line with the requirements of the middle-term plan.

## 4. THE DATA AND THE IMPLEMENTATION

The data used for the application of the artificial neural networks approach for budget expenditure forecasting is given in Table 1. The first half of Table 1 presents the budget expenditure allowances and realizations for both public institutions, while the second half includes the changes in the budget expenditure allowances and expenditures, and the inflation rates for the period of 1973-2008. The trend of changes in the budget expenditure allowances and realizations varies during the period. Usually, the increases in expenditures exceed the increases in inflation rates for both public institutions, and this trend is stable for the period after 2002. There are decreases in both allowances and expenditures in 1973, 1975, 1979 and 1983. With the exception of 1976 the decreases is negligible. This creates a good base for the application of the artificial neural networks approach on the data of the SPO and the CoA.

The expenditures of 1973-2008 of the SPO and the CoA time series are forecasted by using the feed forward neural networks. Each series includes annually 36 observations. The last 5 observations corresponding the years between 2004 and 2008 are used for test set and the rest of the observations are employed for training. Both the number of inputs and neurons in the hidden layer are changed from 1 to 12. Thus, 144 architectures are totally examined for each time series since one neuron is used in the output layer. Then, the best architecture that has the minimum root mean square error (RMSE) value calculated over the test set is selected. In other words, the RMSE is preferred as performance measure since it is a well known measure for forecasting.

**Table 1:** The allowances and realizations of the CoA and the SPO, and the inflation rates (*IR*) (million TL[b])

| | the CoA | | the SPO | | the CoA | | the SPO | | |
|---|---|---|---|---|---|---|---|---|---|
| Years | Allovances[*] | Expenditures | Allovances[**] | Expenditures | Increase (%) | Increase (%) | Increase (%) | Increase (%) | IR[a] |
| 1973 | 40.80 | 52.90 | 152.00 | 118.80 | --- | --- | --- | --- | 21.0 |
| 1974 | 56.40 | 51.90 | 161.40 | 98.30 | 38.24 | -1.89 | 6.18 | -17.26 | 26.9 |
| 1975 | 70.60 | 104.00 | 196.70 | 139.70 | 25.18 | 100.39 | 21.87 | 42.12 | 11.4 |
| 1976 | 126.60 | 99.90 | 74.60 | 67.60 | 79.32 | -3.94 | -62.07 | -51.61 | 17.3 |
| 1977 | 131.10 | 141.00 | 79.50 | 92.50 | 3.55 | 41.14 | 6.57 | 36.83 | 28.5 |
| 1978 | 145.80 | 198.00 | 228.60 | 147.20 | 11.21 | 40.43 | 187.55 | 59.14 | 53.6 |
| 1979 | 294.10 | 326.00 | 227.70 | 227.80 | 101.71 | 64.65 | -0.39 | 54.76 | 75.1 |
| 1980 | 460.80 | 533.30 | 1463.50 | 576.90 | 56.68 | 63.59 | 542.73 | 153.25 | 90.3 |
| 1981 | 698.90 | 595.10 | 1573.00 | 734.70 | 51.67 | 11.59 | 7.48 | 27.35 | 34.1 |
| 1982 | 858.10 | 694.40 | 1675.30 | 863.60 | 22.78 | 16.69 | 6.50 | 17.54 | 27.4 |
| 1983 | 1145.20 | 1305.30 | 2348.90 | 1491.70 | 33.46 | 87.98 | 40.21 | 72.73 | 28.1 |
| 1984 | 1392.10 | 1424.70 | 2749.60 | 1653.30 | 21.56 | 9.15 | 17.06 | 10.83 | 46.4 |
| 1985 | 3528.50 | 3085.40 | 6338.60 | 5098.10 | 153.47 | 116.56 | 130.53 | 208.36 | 41.7 |
| 1986 | 7743.70 | 6515.70 | 13594 | 11634.50 | 38.24 | 111.18 | 114.47 | 128.21 | 27.5 |
| 1987 | 9812.00 | 7455.00 | 25695 | 22961.00 | 26.71 | 14.42 | 89.02 | 97.35 | 39.3 |
| 1988 | 13908.00 | 10715.00 | 31799 | 22753.00 | 38.24 | 43.73 | 23.76 | -0.91 | 60.8 |
| 1989 | 21303.00 | 20100.00 | 46788 | 44035.00 | 53.17 | 87.59 | 47.14 | 93.53 | 65.0 |
| 1990 | 36940.00 | 36648.00 | 78142 | 64033.00 | 73.40 | 82.33 | 67.01 | 45.41 | 49.6 |
| 1991 | 78802.00 | 66471.00 | 128462 | 108804.00 | 113.32 | 81.38 | 64.40 | 69.92 | 52.6 |
| 1992 | 198273.00 | 153768.00 | 248536 | 142891.00 | 151.61 | 131.33 | 93.47 | 31.33 | 67.1 |
| 1993 | 352416.00 | 227728.00 | 240870 | 236510.00 | 77.74 | 48.10 | -3.08 | 65.52 | 55.2 |
| 1994 | 556433.00 | 330780.00 | 431656 | 340438.00 | 57.89 | 45.25 | 79.21 | 43.94 | 120.5 |
| 1995 | 669884.00 | 555780.00 | 646199 | 535397.00 | 20.39 | 68.02 | 49.70 | 57.27 | 88.0 |
| 1996 | 1247105.00 | 1349603.00 | 1459625 | 1306482.00 | 38.24 | 142.83 | 125.88 | 144.02 | 71.6 |
| 1997 | 2805154.50 | 2946822.91 | 2425987 | 2413411.80 | 124.93 | 118.35 | 66.21 | 84.73 | 78.4 |
| 1998 | 5486620.00 | 5333797.86 | 4506500 | 4265065.82 | 95.59 | 81.00 | 85.76 | 76.72 | 68.8 |
| 1999 | 10310357.00 | 10924310.00 | 7843585 | 6708953.00 | 87.92 | 104.81 | 74.05 | 57.30 | 43.7 |
| 2000 | 17535491.00 | 12016914.00 | 13775295 | 10034974.00 | 70.08 | 10.00 | 75.62 | 49.58 | 53.7 |
| 2001 | 22090474.00 | 18997433.00 | 16478750 | 13341840.00 | 25.98 | 58.09 | 19.63 | 32.95 | 57.7 |

*Table 1: cont....*

| 2002 | 40452800.00 | 28717026.12 | 28463000 | 19285766.88 | 83.12 | 51.16 | 72.73 | 44.55 | 45.0 |
| 2003 | 49518000.00 | 32938709.65 | 31971000 | 24341295.15 | 22.41 | 14.70 | 12.32 | 26.21 | 23.8 |
| 2004 | 53598384.00 | 36132894.00 | 78724793 | 49297756.00 | 8.24 | 9.70 | 146.24 | 102.53 | 11.5 |
| 2005 | 54231715.00 | 39814762.15 | 117183597 | 80732234.90 | 1.18 | 10.19 | 48.85 | 63.76 | 5.8 |
| 2006 | 54431000.00 | 48412000.00 | 135847000 | 131750.00 | 0.37 | 21.59 | 15.93 | 63.19 | 7.8 |
| 2007 | 85125000.00 | 62922000.00 | 374283000 | 150328.00 | 56.39 | 29.97 | 175.52 | 14.10 | 8.7 |
| 2008 | 90202350.00 | 76204000.00 | 267070000 | 225849.00 | 5.96 | 21.11 | -28.64 | 50.24 | 11.4 |

[*] and[**] Ex-ante allowances for the period 1973-2008.

[a] Taken from the Istanbul Trade Chamber. (1963=100). The detailed information is also available at the Turkish Statistics Institute (TUIK), [16].

[b] Until 2005 the figures are in Million Turkish Lira (TL). In 2006 six zeros are eliminated from the TL, thus after 2006 the figures are in Thousand TL.

When the best architecture is being searched, the other elements of the artificial neural networks are fixed. The logistic activation function is used in all of the neurons of networks. Levenberg Marquardt algorithm is employed as training algorithm because of the high convergence speed of the algorithm.

The determined best architecture for the time series and the RMSE values produced by these architectures are summarized in Table **2**. According to Table **2**, when the architecture including 3 and 7 neurons in the input and hidden layers, respectively, is used, the best forecasts are obtained for the SPO series and the architecture produces 13523450.88 RMSE value.

**Table 2:** The best architectures for the time series

| Time series | The best architecture | RMSE |
|---|---|---|
| SPO | 3-7-1 | 13523450.88 |
| CoA | 1-4-1 | 1266301.40 |

The obtained RMSE values are not enough to see the performance of the determined artificial neural networks models. Therefore, the obtained results are also examined visually. The graphs of the forecasts obtained from the determined models and the observed values for test sets are given in Figs. **2** and **3**. When the graphs are examined, it is clearly seen that the calculated forecasts are very good.

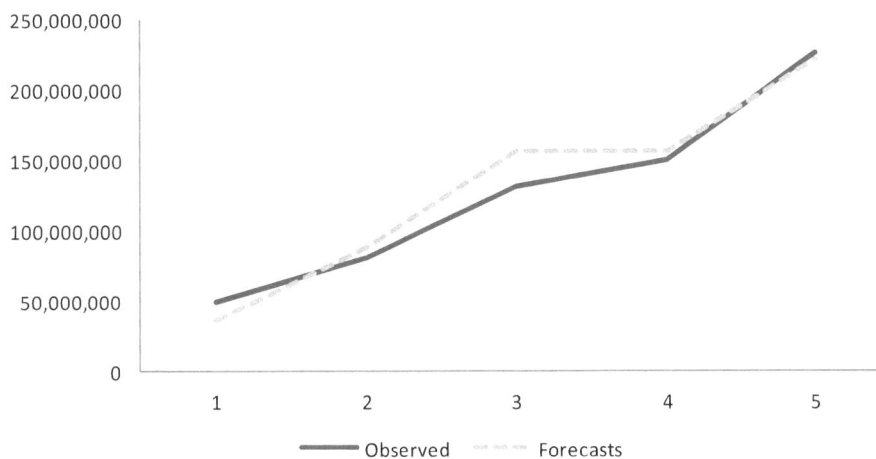

**Figure 2:** The obtained forecasts and the observed values for the SPO series.

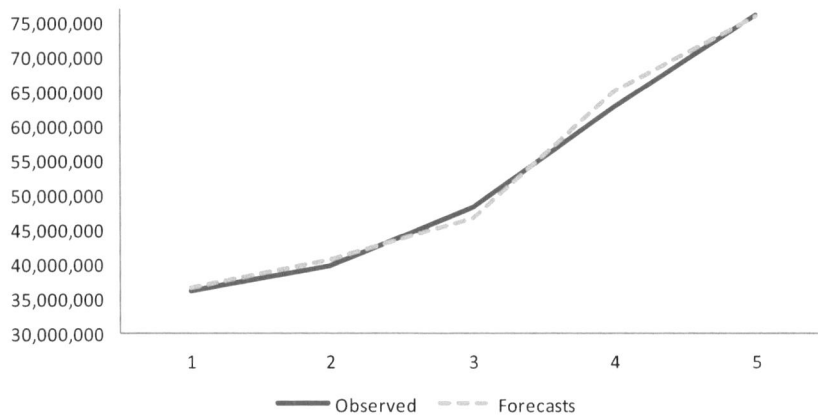

**Figure 3:** The obtained forecasts and the observed values for the CoA series.

## 5. CONCLUSION

In recent years, artificial neural networks have been successfully used for time series forecasting in various fields since this method can model both linear and non linear part of real life time series. On the other hand, neural networks approach has been rarely used for public expenditure forecasting particularly in Turkey. Therefore, in this study, feed forward neural networks are utilized to forecast public expenditures of some institutions in Turkey to obtain accurate forecasts.

Under the new public fiscal management approach in Turkey the public institutions are expected to spend the allocated revenues to meet the realistically determined expenditures effectively, economically and productively. At this point it is important for these public institutions to have a powerful forecasting methodology at their disposal to predict their mid-term expenditures correctly. The implementation on the expenditure data of the SPO and the CoA has proved that the artificial neural networks approach is a very useful prediction tool, which can also be employed by other public institutions to forecast their mid-term expenditures more accurately.

## REFERENCES

[1]   C.H. Aladag, "A new architecture selection method based on tabu search for artificial neural networks, *Expert Syst Appl*, vol. 38, pp. 3287-3293, 2011.

[2]   C.H. Aladag, E. Eğrioğlu, and C. Kadilar, "Forecasting nonlinear time series with a hybrid methodology, *Appl Math Lett*, vol. 22, pp. 1467-1470, 2009.

[3]   M. Bagdigen, "An empirical analysis of accurate budget forecasting in Turkey," *Dogus Univ J*, vol. 6, pp. 190-201, 2005.

[4]   Y. Akyüz, and K. Boratav, "The making of the Turkish financial crisis," *World Dev*, vol. 31, no. 9, pp. 1549-1566, 2003.

[5]   Y.C. Tang, "An approach to budget allocation for an aerospace company-Fuzzy analytic hierarchy process and artificial neural network," *Neurocomputing*, vol. 72, pp. 3477-3489, 2009.

[6]   Baker, B.D., Richards, C.E., A comparison of conventional linear regression methods and neural networks for forecasting educational spending, *Econ Educ Rev*, vol. 18, 405-415, 1999.

[7]   Abeysinghe,T., Jayawickrama, A., Singapore's recurrent budget surplus: The role of conservative growth forecast, *J Asian Econ,* vol. 19, 117-124, 2008.

[8]   C.H. Aladag, E. Eğrioğlu, and S. Gunay, "A new architecture selection strategy in solving seasonal autoregressive time series by artificial neural networks," *Hacet J Math Stat*, vol. 37, no. 2, pp. 185-200, 2008.

[9]   P.D. Picton, *Introduction to neural networks*, Macmillan Press Ltd., 1994

[10]  S. Gunay, E. Eğrioğlu, and C.H. Aladag, *Introduction to single variable time series analysis*, Hacettepe University Press: Ankara, 2007.

[11]　J.M. Zurada, *Introduction of artificial neural systems*, St. Paul: West Publishing, 1992.

[12]　K.A. Smith, *Neural networks in business: techniques and applications*, Imprint Info Hershey: Idea Group, 2002.

[13]　Cichocki, and R. Unbehauen, R., *Neural networks for optimization and signal processing*, John Willey & Sons: New York, 1993.

[14]　G. Zhang, and B.E. Patuwo, and Y.M. Hu, "Forecasting with artificial neural networks: the state of the art," *Int J Forecasting*, vol. 14, pp. 35-62, 1998.

[15]　M. Aksaraylı, H. Egeli, K. Tüğen, H.A. Akdeniz, and A. Özen, "Stratejik Planlamaya Dayalı Performans Esaslı Bütçeleme Sisteminin Etkinliğinin Belirleyicileri: Türkiye'deki Kamu Kurumlarının Analizi," *Süleyman Demirel Üniversitesi İktisadi ve İdari Bilimler Fakültesi Dergisi*, vol. 12, no. 3, pp. 313-326, 2007. (in Turkish)

[16]　Turkish Statistical Institute, *Statistical Indicators 1923-2008*, Publication Number: 3361, pp. 587-588, 2009.

48          *Advances in Time Series Forecasting*, 2012, 48-55

# CHAPTER 7

# A New Method for Forecasting Fuzzy Time Series with Triangular Fuzzy Number Observations

## Erol Eğrioğlu[1,*], Cagdas Hakan Aladag[2] and Ufuk Yolcu[1]

*[1]Ondokuz Mayis University, Faculty of Arts and Science, Department of Statistics, 55139, Samsun, Turkey and [2]Hacettepe University, Faculty of Science, Department of Statistics, 06800, Ankara, Turkey*

**Abstract:** Most of the time series faced in real life are fuzzy time series and these time series have to be forecasted by fuzzy time series forecasting methods. Therefore, there have been many studies in the literature in which various fuzzy time series approaches are proposed. The fuzzy time series methods introduced in the literature have been generally proposed to analyze fuzzy time series whose observations are fuzzy sets. On the other hand, Song et al. firstly improved a fuzzy time series model to analyze fuzzy time series whose observations are triangular fuzzy numbers [1]. Their method requires complex arithmetic operations for triangular fuzzy numbers. We propose a novel fuzzy time series forecasting approach based on simulation and feed forward neural networks to forecast fuzzy time series including triangular fuzzy numbers. The proposed method is applied to gold prices in Turkey series to show the applicability of the method.

**Keywords:** Artificial neural networks, Fuzzy time series, Forecasting, Gold prices in Turkey, Triangular fuzzy number.

## 1. INTRODUCTION

Fuzzy time series are time series whose observations are fuzzy sets. In the literature, various approaches have been proposed and have been used to analyze fuzzy time series. Observations of fuzzy time series can be fuzzy sets and can also be triangular fuzzy numbers which are special type of fuzzy sets. A definition of fuzzy time series based on triangular fuzzy numbers was firstly introduced by Song et al. [1]. Then, Hong [2] defined the arithmetic operations related to the triangular fuzzy numbers given in [1] by using weakest t-norm and extended the definitions presented in [1]. Although there have been many studies in fuzzy time series literature, there have been only these two studies which are based on triangular fuzzy numbers. Furthermore, in these studies, any implementation was not included and it was not explained how the proposed methods work.

The most of the studies in fuzzy time series literature have been focused on the methods that proposed to forecast fuzzy time series whose observations are fuzzy sets consist of intervals. Song and Chissom [3] firstly proposed an approach to forecast such fuzzy time series. Then, Chen proposed new approaches in [4, 5] and these methods are well known ones in the literature. In addition, Tseng [6, 7] proposed new models which are like traditional time series models. In these models, however, the parameters are fuzzy numbers.

In this study, a novel fuzzy time series forecasting model based on simulation and feed forward neural networks is proposed to forecast fuzzy time series whose observations are triangular fuzzy numbers. Using the proposed method easier than using the methods presented in [1, 2] since the proposed method does not require complex arithmetic operations for triangular fuzzy number. Also the proposed approach is applied to gold prices in Turkey series in the implementation.

This chapter gives in Section 2 brief information about artificial neural networks. Section 3 includes the basic definitions for the proposed method and the algorithm of the method is introduced in this section. The implementation is given in Section 4. Finally, Section 5 provides the concluding remarks.

*Address correspondence to Erol Eğrioğlu: Ondokuz Mayis University, Faculty of Arts and Science, Department of Statistics, 55139, Samsun, Turkey; E-mail: erole@omu.edu.tr

## 2. ARTIFICIAL NEURAL NETWORKS

Artificial neural networks are mathematical algorithms that simply mimic the biological neural networks. Although their structure is not as complex as the structure of biological neural networks, many real life problems can be solved by using artificial neural networks. There are different types of artificial neural networks [8]. Especially feed forward neural networks have been preferred in time series forecasting problem since this kind of neural network has been proved its success in many forecasting studies [9].

Feed forward neural networks compose of layers such as input, hidden and output layers. And each layer consists of elements which are called neuron. Layers are connected with each other by weights. However, there is no interconnection with neurons in the same layer. The number of hidden layer can be more than one. In time series forecasting, the input values are the lagged variables of the time series. Target value is the value belongs to the next period and the output of the network will be the prediction value for the next period. A broad multi layer feed forward neural network architecture that contains one hidden layer and one neuron in the output layer is illustrated in Fig. **1**.

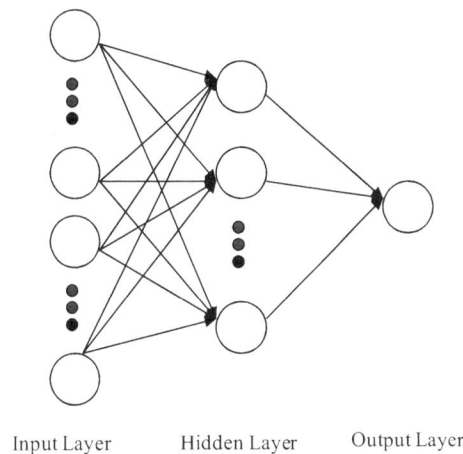

Input Layer          Hidden Layer          Output Layer

**Figure 1:** Multilayer feed forward artificial neural network with one output neuron.

After the inputs for neurons in the hidden and the output layers are multiplied by corresponding weights, they are summed and the value of the activation function is calculated for this sum. By employing activation function, artificial neural networks can model the non linear part of time series. To produce accurate forecasts, artificial neural networks can be trained. In the training process, the best values of the weights are tried to be found. The closer the difference between the prediction and the target value is, the better the weight values are. To train artificial neural networks, various optimization algorithms are used and these algorithms are called as training algorithms [10].

Feed forward neural networks have produced very good results in time series forecasting applications. On the other hand, to reach accurate forecasts, the elements of artificial neural networks such as architecture structure, activation function and training algorithm should be carefully determined since these main components have important effect on forecasting performance of the method [11].

## 3. THE PROPOSED METHOD

In recent years, various approaches have been proposed to forecast fuzzy time series. The observations of fuzzy time series can be fuzzy sets or fuzzy numbers, which can be considered as special type of fuzzy sets. Most of fuzzy time series forecasting approaches available in the literature were improved to forecast fuzzy time series whose observations consist of elements that are subsets of universe of discourse. On the other hand, in the studies [1, 2], two forecasting approaches were proposed to forecast fuzzy time series consist of observations that are fuzzy numbers. These two approaches are based on arithmetic operations for fuzzy numbers. In this study, we propose a novel forecasting model based on feed forward neural networks and simulation in order to forecast fuzzy time series whose observations are triangular fuzzy numbers. The proposed method can be used for time series such as share prices of stockholders and gold prices which have lowest and highest values for a specific time interval. Besides, it is possible to apply the proposed method to other time series which can be considered as fuzzy time series. The fuzzy time series forecasting approach proposed in this study provides some advantages. And, these advantages can be summarized as follows:

- Using the proposed method is easier than using the methods introduced in [1, 2] since the proposed method does not require complex arithmetic operations related to fuzzy numbers.

- The proposed method does not require satisfying any assumptions such as linearity, normal distribution and a specific observation number.

- The proposed method has the ability of flexible modeling which is also included by artificial neural networks.

- The proposed approach produces more accurate forecasts than those obtained from most of the methods available in the literature.

The algorithm of the proposed forecasting method is given below:

**Algorithm of the Proposed Method**

*Step 1. Fuzzification*

The fuzzification process can be changed depend on the observed time series. If the time series has the lowest and the highest values in time intervals in which the time series observations are observed, in the fuzzification step, the lowest and the highest values are taken as the left and the right side values of the triangular fuzzy number for each observations, respectively. The center value of the triangular fuzzy number is equal to mean value of the corresponding left and right side values. In this way, each observation is turned into a triangular fuzzy number. If there is only a crisp value for the time interval in which the corresponding observation of the time series are observed, this crisp value is taken as the center value of the triangular fuzzy number. Then, a spread value is determined and the left and the right side values are calculated by using this spread value. Therefore, the obtained triangular fuzzy number will be symmetrical one.

*Step 2. Determining Fuzzy Relations*

To establish fuzzy relationships, feed forward neural networks are used. By using simulation method, the fuzzy time series whose observations are triangular fuzzy numbers is turned into a time series composes of crisp observations in order to apply feed forward neural networks. A fuzzy time series $F(t)$ can be given as follows:

$$F(t) = (l(t),\ m(t),\ r(t)),\ t = 1,2,\ldots,n \tag{1}$$

where $l(t), m(t)$ and $r(t)$ are the left side, the center and the right side value, respectively, for the $i$ th observation of the fuzzy time series.

*Step 2.1. Calculating the Crisp Time Series*

First of all, it is determined how many crisp time series will be analyzed and this number can be represented by itr. When each crisp time series is generated, the following way is used.

- It is assumed that $\alpha$ has a uniform distribution ($\alpha \sim \mathrm{Unifrom}(0,1)$) so an $\alpha$ value can be generated by using this distribution.

- The obtained $\alpha$ cut values for every observation can be calculated by using the equation given in (2) depend on the value of that $\alpha$.

$$F(t)[\alpha] = \left[\, l(t) + \left(m(t) - l(t)\right) * \alpha, r(t) - \left(r(t) - m(t)\right) * \alpha\,\right], t = 1, 2, \ldots, n \qquad (2)$$

- A crisp value for each observation is calculated by using uniform distribution whose parameters are the values of the lower and upper bounds of the corresponding $\alpha$ cut value. Thus, time series with $n$ observations given below are obtained. Total number of these time series is equal to itr as mentioned before.

$$x_j(t) \sim \mathbf{Uniform}\left[\, l_j(t) + \left(m_j(t) - l_j(t)\right) * \alpha, r_j(t) - \left(r_j(t) - m_j(t)\right) * \alpha\,\right],$$

$$t = 1,2,\ldots,n, \; j = 1,2,\ldots,itr$$

### Step 2.2. Analyzing the crisp time series by using artificial neural networks

The crisp time series obtained in the previous step ($x_j(t)$, $t=1,2,\ldots,n$, $j=1,2,\ldots,itr$) are analyzed with feed forward neural networks. The input values for feed forward neural networks models are the lagged variables of time series $x_j(t)$. The time series is divided into two sets which are used for training and test. After the best weight values are found using the training set, the performance of the feed forward artificial neural network model is measured using the test set. By using the best artificial neural network model that gives the best performance for test set, predictions are calculated for the training and the test sets. In other words, $\hat{x}_j(t)$, ($t = 1,2,\ldots,n$, $j = 1,2,\ldots,itr$) time series are obtained.

### Step 3. Calculating the Fuzzy Forecasts

For each observation, predictions for the lowest, the center and the highest values are calculated from the obtained predicted time series. These calculated predictions for each observation are the left side, the center and the right side values, respectively, for the corresponding fuzzy forecast. Fuzzy forecasts can be given as follows:

$$\widehat{F}(t) = \left(\min_j\left(x_j(t)\right), \operatorname*{mean}_j\left(x_j(t)\right), \max_j\left(x_j(t)\right)\right), t = 1, 2, \ldots, n$$

### Step 4. Calculating the Defuzzyfied Forecasts

It is not necessary to perform the defuzzification step. It is optional. If each observation of the original data has the lowest and the highest values, the left side and the right side values of the obtained fuzzy forecasts are the prediction values for the lowest and the highest values of the corresponding observations, respectively. On the other hand, if every observation in the original time series has only one real value, the defuzzification step is needed. For such time series, the centroid method is used in the defuzzification step. In this case, the defuzzyfied forecasts will be the center values of the fuzzy forecasts obtained in the previous step.

## 4. THE IMPLEMENTATION

To show the applicability of the proposed method, it is applied to the lowest and the highest gold prices in Turkey series whose observations are between 27.01.2009 and 30.04.2009. The daily time series are shown in Table **1**.

In the fuzzification step of the proposed method, the lowest value, the highest value and the mean value of these values were taken as the left side, the right side and the center values of the triangular fuzzy number for each observation. In Step 2.1, 500 crisp time series were generated so itr = 500. The number of inputs of feed forward neural networks, which is also equal to the order of the model, and the number of neurons in the hidden layer were varied between 1 and 5. Thus, 25 architectures were totally examined to find the best architecture. In the architecture selection process, the time series was divided into two sets which are training and test sets. The training set included the daily observations between 27.01.2009 and 16.09.2009 and the test set consisted of the observations from 17.04.2009 to 30.04.2009.

**Table 1:** The time series of the lowest and the highest gold prices

| Date | The Lowest Gold Price | The Highest Gold Price | Date | The Lowest Gold Price | The Highest Gold Price |
|------|------|------|------|------|------|
| 27.01.2009 | 46580 | 46760 | 16.03.2009 | 50100 | 50580 |
| 28.01.2009 | 45930 | 46360 | 17.03.2009 | 50050 | 50385 |
| 29.01.2009 | 45570 | 45840 | 18.03.2009 | 49970 | 49985 |
| 30.01.2009 | 47850 | 48400 | 19.03.2009 | 50530 | 51250 |
| 02.02.2009 | 48100 | 48480 | 20.03.2009 | 51430 | 52850 |
| 03.02.2009 | 47650 | 47780 | 23.03.2009 | 50740 | 51100 |
| 04.02.2009 | 47200 | 47275 | 24.03.2009 | 49070 | 49785 |
| 05.02.2009 | 47680 | 48277 | 25.03.2009 | 49220 | 49700 |
| 06.02.2009 | 47490 | 48030 | 26.03.2009 | 49710 | 49900 |
| 09.02.2009 | 46534 | 46840 | 27.03.2009 | 49550 | 49800 |
| 10.02.2009 | 46400 | 46980 | 30.03.2009 | 49600 | 50190 |
| 11.02.2009 | 47970 | 48440 | 31.03.2009 | 49260 | 49725 |
| 12.02.2009 | 49609 | 49715 | 01.04.2009 | 49000 | 49195 |
| 13.02.2009 | 48995 | 49585 | 02.04.2009 | 47100 | 48150 |
| 16.02.2009 | 49750 | 50200 | 03.04.2009 | 46170 | 46355 |
| 17.02.2009 | 51375 | 52200 | 06.04.2009 | 44100 | 44475 |
| 18.02.2009 | 52140 | 53000 | 07.04.2009 | 44800 | 45550 |
| 19.02.2009 | 52350 | 52475 | 08.04.2009 | 45300 | 46150 |
| 20.02.2009 | 53245 | 54460 | 09.04.2009 | 44345 | 44950 |
| 23.02.2009 | 53080 | 53900 | 10.04.2009 | 44500 | 44650 |
| 24.02.2009 | 53560 | 53700 | 13.04.2009 | 44860 | 45300 |
| 25.02.2009 | 51445 | 52060 | 14.04.2009 | 45000 | 45445 |
| 26.02.2009 | 50985 | 51780 | 15.04.2009 | 45400 | 45985 |
| 27.02.2009 | 51065 | 52210 | 16.04.2009 | 45800 | 45875 |
| 02.03.2009 | 52260 | 52950 | 17.04.2009 | 45110 | 45220 |
| 03.03.2009 | 50780 | 51225 | 20.04.2009 | 45255 | 46193 |
| 04.03.2009 | 50605 | 50870 | 21.04.2009 | 47300 | 47610 |
| 05.03.2009 | 50650 | 51485 | 22.04.2009 | 46733.75 | 47283.25 |
| 06.03.2009 | 53450 | 53930 | 23.04.2009 | 46733.75 | 47283.25 |
| 09.03.2009 | 53650 | 54530 | 24.04.2009 | 47090 | 47520 |
| 10.03.2009 | 51590 | 52400 | 27.04.2009 | 47290 | 47810 |
| 11.03.2009 | 49920 | 51000 | 28.04.2009 | 46620 | 47235 |
| 12.03.2009 | 50930 | 51955 | 29.04.2009 | 46180 | 46320 |
| 13.03.2009 | 50060 | 51200 | 30.04.2009 | 45365 | 45890 |

After the proposed method was applied to the lowest and the highest gold prices time series, the forecasts were calculated over the test sets. Three criteria were used to measure the forecasting performances. These measures are root of mean square error (RMSE), mean absolute percentage error (MAPE) and direction accuracy (DA). The related formulas for these criteria are given as follows:

$$RMSE = \left( \frac{\sum_{i=1}^{T}(y_i - \hat{y}_i)^2}{T} \right)^{1/2} \tag{3}$$

$$MAPE = \frac{1}{T}\sum_{i=1}^{T}\left|\frac{y_i - \hat{y}_i}{y_i}\right| \tag{4}$$

$$DA = \frac{1}{T}\sum_{i=1}^{T}a_i \quad , \quad a_i = \begin{cases} 1 & if \ (y_{i+1} - y_i)(\hat{y}_{i+1} - y_i) > 0 \\ 0 & otherwise. \end{cases} \tag{5}$$

where $y_i$ is the actual value; $\hat{y}_i$ is the predicted value; $T$ is the number of data and $m$ is the number of weights. For all of the examined feed forward neural networks, the measures of these criteria were calculated for both time series. The obtained results are summarized in Tables 2–7.

Table 2: The obtained RMSE values for the lowest prices

| Number of Inputs | Number of Hidden Layer Neurons | | | | |
| | 1 | 2 | 3 | 4 | 5 |
|---|---|---|---|---|---|
| 1 | 673.6446 | 680.3016 | 673.8480 | 585.9328 | 672.3737 |
| 2 | 653.7073 | 690.6383 | 827.5978 | 1151.0971 | 2186.9813 |
| 3 | 655.6084 | 1055.8088 | 716.7808 | 1180.3612 | 942.6865 |
| 4 | 630.5884 | 1009.6077 | 1150.5045 | 2255.8410 | 2246.6066 |
| 5 | 783.6248 | 1031.7266 | 751.8096 | 2129.0855 | 2198.3972 |

Table 3: The obtained MAPE values for the lowest prices

| Number of Inputs | Number of Hidden Layer Neurons | | | | |
| | 1 | 2 | 3 | 4 | 5 |
|---|---|---|---|---|---|
| 1 | 0.0117 | 0.0114 | 0.0116 | 0.0084 | 0.0119 |
| 2 | 0.0112 | 0.0121 | 0.0137 | 0.0187 | 0.0436 |
| 3 | 0.0115 | 0.0154 | 0.0125 | 0.0208 | 0.0153 |
| 4 | 0.0112 | 0.0156 | 0.0195 | 0.0452 | 0.0450 |
| 5 | 0.0145 | 0.0166 | 0.0131 | 0.0428 | 0.0436 |

Table 4: The obtained DA values for the lowest prices

| Number of Inputs | Number of Hidden Layer Neurons | | | | |
| | 1 | 2 | 3 | 4 | 5 |
|---|---|---|---|---|---|
| 1 | 0.6667 | 0.6667 | 0.6667 | 0.6667 | 0.4444 |
| 2 | 0.6667 | 0.4444 | 0.5556 | 0.4444 | 0.4444 |
| 3 | 0.7778 | 0.6667 | 0.5556 | 0.3333 | 0.5556 |
| 4 | 0.6667 | 0.5556 | 0.5556 | 0.4444 | 0.4444 |
| 5 | 0.6667 | 0.7778 | 0.6667 | 0.4444 | 0.4444 |

Table 5: The obtained RMSE values for the highest prices

| Number of Inputs | Number of Hidden Layer Neurons | | | | |
| | 1 | 2 | 3 | 4 | 5 |
|---|---|---|---|---|---|
| 1 | 712.8913 | 698.4937 | 697.6451 | 670.8139 | 784.5251 |
| 2 | 671.7747 | 689.7105 | 957.2230 | 2371.4428 | 4520.8780 |
| 3 | 750.9600 | 2730.6173 | 739.1233 | 1690.9773 | 5596.5286 |
| 4 | 751.7648 | 764.2048 | 2875.4425 | 5244.9302 | 6871.6598 |
| 5 | 805.0305 | 748.9149 | 832.6643 | 6596.9235 | 7174.5700 |

**Table 6:** The obtained MAPE values for the highest prices

|  | Number of Hidden Layer Neurons | | | | |
|---|---|---|---|---|---|
| Number of Inputs | 1 | 2 | 3 | 4 | 5 |
| 1 | 0.0120 | 0.0125 | 0.0120 | 0.0114 | 0.0132 |
| 2 | 0.0123 | 0.0126 | 0.0179 | 0.0358 | 0.0658 |
| 3 | 0.0135 | 0.0559 | 0.0135 | 0.0329 | 0.1063 |
| 4 | 0.0129 | 0.0136 | 0.0402 | 0.0995 | 0.1395 |
| 5 | 0.0139 | 0.0131 | 0.0141 | 0.1372 | 0.1516 |

**Table 7:** The obtained DA values for the highest prices

|  | Number of Hidden Layer Neurons | | | | |
|---|---|---|---|---|---|
| Number of Inputs | 1 | 2 | 3 | 4 | 5 |
| 1 | 0.5556 | 0.4444 | 0.5556 | 0.7778 | 0.5556 |
| 2 | 0.4444 | 0.4444 | 0.4444 | 0.3333 | 0.4444 |
| 3 | 0.4444 | 0.4444 | 0.5556 | 0.4444 | 0.4444 |
| 4 | 0.3333 | 0.3333 | 0.4444 | 0.4444 | 0.4444 |
| 5 | 0.5556 | 0.7778 | 0.5556 | 0.4444 | 0.4444 |

According to Tables **2–7**, for both time series, when one input (first order model) and four neurons in the hidden layer are used in the architectures, the most accurate results are obtained in terms of RMSE and MAPE criterion. Also, this architecture, which can be represented by 1-4-1, produces the best forecasts for both time series in terms of DA criterion.

To make a comparison, the lowest and the highest gold prices series were also forecasted by using the methods introduced by Chen in [4, 5]. When these time series were forecasted by employing the methods proposed in [4, 5], the lengths of interval are 1360 and 1430 for the lowest and the highest prices time series, respectively. The forecasting performance measures values obtained from these two methods are presented in Table **8**.

**Table 8:** The Results obtained from the methods given in [4] and [5]

| Time Series | Criteria | Chen [4] | Chen [5] | | | |
|---|---|---|---|---|---|---|
|  |  |  | 2. Order | 3. Order | 4. Order | 5. Order |
|  | RMSE | 1169.0 | 1177.0 | 1181.0 | 1093.0 | 1027.2 |
| The Lowest Price | MAPE | 0.0209 | 0.0181 | 0.0180 | 0.0169 | 0.0153 |
|  | DA | 0.6667 | 0.4400 | 0.5500 | 0.5500 | 0.5556 |
|  | RMSE | 818.19 | 1022.00 | 970.36 | 925.48 | 916.26 |
| The Highest Price | MAPE | 0.0142 | 0.0170 | 0.0163 | 0.0150 | 0.0142 |
|  | DA | 0.6667 | 0.4400 | 0.5556 | 0.5556 | 0.5556 |

As it is seen from Table **8**, for the lowest gold price series, the best forecasts are obtained by using the method introduced in [5] when the model order is 5 in terms of RMSE and MAPE criteria. For the highest gold price series, the most accurate results are obtained when the method proposed by Chen [4] is employed in terms of RMSE and MAPE criteria. It can also be said that the method proposed by Chen [5] produces very good forecasts for this time series when the model order is 5 in terms of MAPE criterion. According to DA criterion, the method introduced by Chen [4] gives the best results for both time series.

Finally, when all of the results obtained from the proposed method and the methods introduced in [4] and [5] are examined, it is clearly seen that the forecasting approach proposed in this study produces the most accurate forecasts in terms of all criteria.

# 5. CONCLUSION

In this study, a new forecasting approach is proposed to forecast fuzzy time series consist of observations which are triangular fuzzy numbers. The proposed method is based on feed forward neural networks and simulation in order to reach high accuracy level. The proposed approach also provides some advantages. It is easy to use the proposed method since it does not include arithmetic operations for fuzzy numbers unlike the methods given in [4, 5]. Another advantage of the proposed method is that it does not include assumption constraints such as normal distribution and linearity. In addition, the proposed model has the flexible modeling ability since feed forward neural networks approach is employed in the method.

To show the applicability of the proposed method, the gold prices in Turkey series were forecasted by using the proposed forecasting approach in the implementation. For the aim of comparison, the time series were also analyzed by using the other methods available in the literature. As a result of the comparison, it was clearly seen that the most accurate forecasts are obtained when the proposed approach is used in terms of RMSE, MAPE and DA criteria.

# REFERENCES

[1]    Q. Song, R.P. Leland, and B.S. Chissom, "A new fuzzy time-series model of fuzzy number observations," *Fuzzy Set Syst*, vol. 73, pp. 341-348, 1995.
[2]    D.H. Hong, "A note on fuzzy time-series model," *Fuzzy Set Syst*, vol. 155, pp. 309-316, 2005.
[3]    Q. Song, and B.S. Chissom, "Forecasting enrollments with fuzzy time series - Part I," *Fuzzy Set Syst*, vol. 54, pp. 1-10, 1993.
[4]    S.M. Chen, "Forecasting enrollments based on fuzzy time-series," *Fuzzy Set Syst*, vol. 81, pp. 311-319, 1996.
[5]    S.M. Chen, "Forecasting enrollments based on high order fuzzy time series," *Cybernet Syst*, vol. 33, pp. 1-16, 2002.
[6]    F.M. Tseng, G.H. Tzeng, and H.C. Yu, "Fuzzy seasonal time series for forecasting the production value of the mechanical industry in Taiwan," *Technol Forecast Soc*, vol. 60, pp. 263-273, 1999.
[7]    F.M. Tseng, G.H. Tzeng, H.C. Yu, and B.J.C. Yuan, "Fuzzy ARIMA model for forecasting the foreign Exchange market," *Fuzzy Set Syst*, vol. 118, pp. 9-19, 2011.
[8]    C.H. Aladag, E. Eğrioğlu, and C. Kadilar, "Forecasting nonlinear time series with a hybrid methodology, *Appl Math Lett*, vol. 22, pp. 1467-1470, 2009.
[9]    C.H. Aladag, E. Eğrioğlu, S. Gunay, and M.A. Basaran, "Improving weighted information criterion by using optimization," *J Comput Appl Math*, vol. 233, pp. 2683-2687, 2010.
[10]   E. Eğrioğlu, C.H. Aladag, and S. Gunay, "A new model selection strategy in artificial neural network" *Appl Math Comput*, vol. 195, pp. 591-597, 2008.
[11]   C.H. Aladag, "A new architecture selection method based on tabu search for artificial neural networks, Expert Syst Appl, vol. 38, pp. 3287-3293, 2011.

# CHAPTER 8

# New Criteria to Compare Interval Estimates in Fuzzy Time Series Methods

## Erol Eğrioğlu[*], V. Rezan Uslu and Senem Koc

*Ondokuz Mayis University, Faculty of Arts and Science, Department of Statistics, 55139, Samsun, Turkey*

**Abstract:** The idea of exploring fuzzy set theory to time series forecasting issues has been enormously attracted researcher's attention in recent years. Several new approaches on fuzzy time series have been put forward. These approaches have got some advantages related to classical methods and are complementary of them. Two of these kinds of procedures are FARIMA and FSARIMA. FARIMA and FSARIMA do not require a restriction of at least 50 observations and linearity assumption. The methods of FARIMA and FSARIMA provide interval estimates of a time series. ARIMA and SARIMA also provide interval estimation but it has been put forward that estimated intervals are large, therefore not informative. The width of estimated intervals obtained from FARIMA and SARIMA may generally tend to be less than ones from ARIMA and SARIMA. In the literature, there has been no study which provides a criterion for the comparisons of time series with respect to interval estimates. In this study, two criteria for such comparisons are presented.

**Keywords:** ARIMA, Fuzzy ARIMA, Fuzzy SARIMA, Interval estimates, SARIMA.

## 1. INTRODUCTION

In time series analysis the linear and nonlinear procedures are generally classified into two groups; univariate and bivariate time series. In time series analysis, the Box–Jenkins methodology, applies autoregressive moving average (ARMA) or autoregressive integrated moving average (ARIMA) models to find the best fit of a time series to past values of this time series, in order to make forecasts. In order to apply the Box-Jenkins methodology, however, two assumptions are generally required; one of them is that the applied model must be linear and the other is that there must be at least 50 observations. Most often researches have been faced some difficulties in providing these assumptions dealing with real life time series data. In that point fuzzy time series approaches can be cure when they face such problems. Since these approaches are data- based approaches there might not need some kind of limitations. Fuzzy logic concept was firstly introduced by [1]. The first implementation of fuzzy logic on time series was explored in the study of [2], by introducing a fuzzy linear regression model. This implementation was used by [3] who puts forward fuzzy ARIMA.

Fuzzy ARIMA and fuzzy SARIMA are aimed to get interval estimates. Traditional ARIMA and SARIMA can provide the point estimation whereas they also provide interval estimates when the normality assumption is maintained. Fuzzy approaches do not require the normality assumption, moreover since the fuzzy interval estimates are generally narrower than the traditional ones they can interpret more easily. In fuzzy literature the comparisons are being made graphically. So far there are no criteria for the comparisons of interval estimates. In this study two new criteria have been proposed to do this. This is shown empirically on the data of the amount of carbon dioxide measured monthly in Ankara. SARIMA and Fuzzy SARIMA have been applied to this data set. For this analysis some subroutines are programmed in MATLAB.

In the second section, fuzzy ARIMA method is summarized. Fuzzy SARIMA method is briefly represented in the third section. In the fourth section, the proposed criteria are introduced. Application of real life time series is given section 5. The obtained results of application are discussed and the conclusions are given in the last section.

---

**\*Address correspondence to Erol Eğrioğlu:** Ondokuz Mayis University, Faculty of Arts and Science, Department of Statistics, 55139, Samsun, Turkey; E-mail: erole@omu.edu.tr

## 2. FUZZY ARIMA

ARIMA is a most powerful analyzing method for time series as long as its assumptions, such as linearity and the number of observations, are satisfied. Most real life time series data cannot, however, provide these assumptions. In ARIMA model, the value of observation at time t of a time series is a linear combination of the value of the observation at time t-1, t-2 ... and error term [4]. In practice it is very difficult to construction such a relationship. [5] proposed Fuzzy SARIMA by exploring Tanaka's possibility approach to Seasonal ARIMA. [3] provides a experimental study of analyzing of a real life time series data *via* Fuzzy ARIMA. The method in this experimental study is summarized as follows:

In Fuzzy ARIMA approach the model parameters are assumed fuzzy as Tanaka's fuzzy regression model. The parameters $\tilde{\varphi} = \left( \tilde{\varphi}_1, ..., \tilde{\varphi}_p \right)$ and $\tilde{\theta} = \left( \tilde{\theta}_1, ..., \tilde{\theta}_q \right)$ are used instead of $\varphi = \left( \varphi_1, ..., \varphi_p \right)$ and $\theta = \left( \theta_1, ..., \theta_q \right)$ in fuzzy ARIMA model, respectively. Then the model is written as

$$\tilde{Z}_t = \tilde{\varphi}_1 Z_{t-1} + \cdots + \tilde{\varphi}_p Z_{t-p} + a_t - \tilde{\theta}_1 a_{t-1} - \cdots - \tilde{\theta}_q a_{t-q} \tag{1}$$

For the simpler notation the model is rewritten as

$$\tilde{Z}_t = \tilde{\beta}_1 Z_{t-1} + \cdots + \tilde{\beta}_p Z_{t-p} + a_t - \tilde{\beta}_{p+1} a_{t-1} - \cdots - \tilde{\beta}_{p+q} a_{t-q} \tag{2}$$

The membership functions for the fuzzy parameters, which are expressed in the form of triangular fuzzy number, are as below.

$$\mu_{(\tilde{\beta}_i)}(\beta_i) = \begin{cases} 1 - \dfrac{|\beta_i - \alpha_i|}{c_i} & , \quad \alpha_i - c_i \leq \beta_i \leq \alpha_i + c_i \\ 0 & , \quad o.w \end{cases} \tag{3}$$

With respect to the extension principal, the membership function of $\tilde{Z}_i$ is

$$\mu_Z(Z_t) = \begin{cases} 1 - \dfrac{\left| Z_t - \sum_{i=1}^p \alpha_i Z_{t-i} - a_t + \sum_{i=p+1}^{p+q} \alpha_i a_{t+p-i} \right|}{\sum_1^p c_i |Z_{t-i}| + \sum_{i=p+1}^{p+q} c_i |a_{t+p-i}|} & , \quad Z_t \neq 0, \ a_t \neq 0, \\ 0 & , \quad o.w. \end{cases} \tag{4}$$

Similar to fuzzy regression, the membership degree of each observation $Z_i$ should be greater than a predetermined value h (h ∈ [0,1]). The choice of h is influenced on the extension of scale parameters. The case is expressed by

$$\mu_Z(Z_t) \geq h, \ \ t = 1, 2, \ldots, k \tag{5}$$

The total uncertainty in fuzzy ARIMA is given by

$$S = \sum_{i=1}^p \sum_{t=1}^k c_i |\phi_{ii}| |Z_{t-i}| + \sum_{i=p+1}^{p+q} \sum_{t=1}^k c_i |\rho_{i-p}| |a_{t+p-i}| \tag{6}$$

where $\varphi_{ii}$ is the ith partial autocorrelation coefficient and $\rho_i$ is the ith autocorrelation coefficient. The estimates of fuzzy ARIMA model parameters are obtained by solving the following linear programming model.

$$Min \quad S = \sum_{i=1}^{p}\sum_{t=1}^{k} c_i |\varphi_{ii}||Z_{t-i}| + \sum_{i=p+1}^{p+q}\sum_{t=1}^{k} c_i |\rho_{i-p}||a_{t+p-i}|$$

YŞ;

$$\sum_{i=1}^{p}\alpha_i Z_{t-i} + a_t - \sum_{i=p+1}^{p+q}\alpha_i a_{t+p-i} + (1-h)\left(\sum_{i=1}^{p} c_i |Z_{t-i}| + \sum_{i=p+1}^{p+q} c_i |a_{t+p-i}|\right) \geq Z_t \quad ,t=1,2,\ldots,k \qquad (7)$$

$$\sum_{i=1}^{p}\alpha_i Z_{t-i} + a_t - \sum_{i=p+1}^{p+q}\alpha_i a_{t+p-i} - (1-h)\left(\sum_{i=1}^{p} c_i |Z_{t-i}| + \sum_{i=p+1}^{p+q} c_i |a_{t+p-i}|\right) \leq Z_t \quad ,t=1,2,\ldots,k$$

$$c_i \geq 0$$

[3] used the three stage procedure, which is given in below, for the application of fuzzy ARIMA.

***Stage 1.*** When the time series is not fuzzy time series ARIMA(p, d, q) model is analyzed by Box-Jenkins procedure. The obtained estimates are taken as the centered value of the fuzzy parameters.

$$\mathbf{\alpha}^* = \left(\alpha_1^*,\ldots,\alpha_{p+q}^*\right) = \left(\hat{\varphi}_1,\ldots,\hat{\varphi}_p,\hat{\theta}_1,\ldots,\hat{\theta}_q\right)$$

***Stage 2.*** The centered values $\mathbf{\alpha}^* = \left(\alpha_1^*,\ldots,\alpha_{p+q}^*\right)$ and the residuals ($a_t$) from Stage 1 are used as the inputs of the problem of minimization. Then it is solved and found the estimates of scale parameters. The fuzzy ARIMA model

$$\tilde{Z}_t = <\alpha_1,c_1>Z_{t-1}+\cdots+<\alpha_p,c_p>Z_{t-p}+a_t-\cdots-<\alpha_{p+q},c_{p+q}>a_{t-q} \qquad (8)$$

***Stage 3.*** The possibility fuzzy regression method is sensitive to outliers then the observations, which are close to the estimates of lower and upper limits of fuzzy ARIMA, are deleted and the fuzzy ARIMA is repeated.

## 3. FUZZY SARIMA

There are some limitations as well as fuzzy ARIMA. At least 50 observations are at hand. In fuzzy SARIMA approach it is also assumed that the parameters are fuzzy. $\tilde{\varphi} = \left(\tilde{\varphi}_1,\ldots,\tilde{\varphi}_p\right)$, $\tilde{\Phi} = (\tilde{\Phi}_1,\ldots,\tilde{\Phi}_p)$, $\tilde{\Theta} = (\tilde{\Theta}_1,\ldots,\tilde{\Theta}_p)$ and $\tilde{\theta} = \left(\tilde{\theta}_1,\ldots,\tilde{\theta}_q\right)$ $\varphi = \left(\varphi_1,\ldots,\varphi_p\right)$ are also formed as triangular fuzzy number. The parameter of $a_t$ is $\tilde{\gamma}_t$. Then the fuzzy SARIMA model with fuzzy parameters ( $p,d,q)(P,D,Q)$ is

$$\tilde{\phi}(B)\tilde{\Phi}(B^s)W_t = \tilde{\beta}_0 + \tilde{\theta}(B)\tilde{\Theta}(B^s)a_t \quad W_t = (1-B)^d(1-B^s)^D Z_t$$

$$\tilde{W}_t = \tilde{\beta}_0 + \sum_{i=1}^{p}\tilde{\phi}_i W_{t-i} + \sum_{i=1}^{p}\tilde{\Phi}_i W_{t-is} - \sum_{i=1}^{p}\phi_i \Phi_1 W_{t-s-i} - \sum_{i=1}^{p}\phi_i \Phi_2 W_{t-2s-i} - \ldots$$

$$-\sum_{i=1}^{p}\phi_i \Phi_p W_{t-Ps-i} + \tilde{\gamma}a_t - \sum_{i=1}^{q}\tilde{\theta}_i a_{t-i} - \sum_{i=1}^{Q}\Theta_i a_{t-is+}$$

where $Z_t$'s are the observations, The membership degree of $\tilde{\gamma}_t$ is 1. The equation above is rewritten as

$$\tilde{W}_t = \tilde{\beta}_0 + \sum_{i=1}^{p}\tilde{\beta}_i W_{t-i} + \sum_{i=1}^{p}\tilde{\beta}_{p+i}W_{t-is} - \sum_{j=1}^{p}\sum_{i=1}^{p}\tilde{\beta}_i\tilde{\beta}_{p+j}W_{t-js-i} + \tilde{\beta}_{p+P+1}a_t - \sum_{i=1}^{q}\tilde{\beta}_{p+P+1+i}a_{t-i}$$

$$\sum_{i=1}^{Q}\tilde{\beta}_{p+P+q+1-i}a_{t-is} + \sum_{j=1}^{Q}\sum_{i=1}^{q}\tilde{\beta}_{p+P+1+i}\tilde{\beta}_{p+P+q+1+j}a_{t-js-i}.$$

When fuzzy parameters are in the form of triangular form the membership function is

$$\mu_{\tilde{B}_i}(\beta_i) = \begin{cases} 1 - |\alpha_i - \beta_i| / c_i \ if \ \alpha_i - c_i \le \beta_i \le \alpha_i + c_i \\ 0 \end{cases}$$

where $\alpha_i$ and $c_i$ denote, respectively, the centered value of fuzzy numbers and the scale parameter. Due to the extension principal, the following formula for the product of fuzzy numbers can be used.

$$A_i \otimes A_j \cong (c_i c_j, a_i a_j, b_i b_j) \tag{9}$$

where $A_i = (c_i, a_i, b_i)$ and $A_j = (c_j, a_j, b_j)$ are the triangular fuzzy numbers. Then the product fuzzy function $\tilde{\beta}_i \tilde{\beta}_{p+j}$ is given,

$$\tilde{\beta}_i \tilde{\beta}_{p+j} = (c_i c_{p+j}, \alpha_i \alpha_{p+j}, c_i c_{p+j}).$$

Finally the membership function of $W$ is

$$\mu_{\tilde{w}}(W_t) = \begin{cases} 1 - |W_t - E_t| / F_t \ for \ W_t \ne 0, a_t \ne 0 \\ 0 \end{cases}$$

where

$$E_t = \alpha_o + \sum_{i=1}^{p} \alpha_i W_{t-i} + \sum_{i=1}^{P} \alpha_{p+i} W_{t-is} - \sum_{j=1}^{P} \sum_{i=1}^{p} \alpha_i \alpha_{p+j} W_{t-js-i} + a_t - \sum_{i=1}^{q} \alpha_{p+P+i} a_{t-i}$$

$$\sum_{i=1}^{Q} \alpha_{p+P+q+i} a_{t-is} + \sum_{j=1}^{Q} \sum_{i=1}^{q} \alpha_{p+P+i} \tilde{\alpha}_{p+P+j} a_{t-js-i} \tag{10}$$

$$F_t = c_o \sum_{i=1}^{p} c_i |W_{t-i}| + \sum_{i=1}^{P} c_{p+i} |W_{t-is}| + \sum_{j=1}^{P} \sum_{i=1}^{p} c_i c_{p+j} |W_{t-js-i}| + c_{p+P+1} |a_t| + \sum_{i=1}^{q} c_{p+P+1+i} |a_{t-i}|$$

$$+ \sum_{i=1}^{Q} c_{p+P+q+1+i} |a_{t-is}| + \sum_{j=1}^{Q} \sum_{i=1}^{q} c_{p+P+1+1} c_{p+P+q+1+j} |a_{t-js-i}|.$$

The problem is finally formed as follows;

$$Min. \ S = \sum_{t=1}^{k} F_t$$

$$\text{Subject to} \quad \begin{aligned} & E_t + (1-h)F_t \ge W_t, t = 1, 2, ..., k, \\ & E_t - (1-h)F_t \le W_t, t = 1, 2, ..., k, \\ & c_i \ge 0, i = 1, 2, ..., p + P + q + Q + 1 \end{aligned} \tag{11}$$

In linear programming model the number of limitations exceeds the number of observations. This makes SARIMA model complicated. In application of fuzzy SARIMA, we use the following two stages formula so that the number of limitations is equal to the number of observations.

***Stage 1.*** The parameters of Fuzzy SARIMA (FSARIMA) model are defined by the traditional procedures. The time series which is not fuzzy is analyzed by Box-Jenkins. The obtained estimates of parameters are taken as the centered value of fuzzy parameters.

$$\alpha^* = \alpha_1, \alpha_2, ..., \alpha_{p+P+q+Q}$$

**Stage 2.** The outputs obtained from the stage 1 are used to solve the minimization problem in (11). The estimation procedure produces interval estimates. The model is $W_t = E_t \pm F_t$

If $c_j$ is very wide then the approach proposed in Stage 3 of Fuzzy ARIMA is highly recommended.

## 4. NEW CRITERIA FOR COMPARISONS OF INTERVAL ESTIMATES

The analyzing of ARIMA and SARIMA model produces point estimations. Contrarily Fuzzy ARIMA and SARIMA aim to provide interval estimates for the model parameters, which are fuzzy number. When the normality assumption is valid then the analyzing of ARIMA and SARIMA model can provide interval estimates. These estimates are generally very wide so they are not very informative. Researchers claim that the interval estimates from FARIMA and SFARIMA are more informative that is narrower. The limits of the interval estimates are considered as the possible values for the minimum and the maximum of the predicted value. The comparisons of both interval estimates from traditional and fuzzy approaches are mostly done by graphically so far. In this study additional to graphical approaches the new criteria are recommended to do this. These are; the range of interval estimates, the number of the interval estimates which include the true value or not. Regarding to these criteria two new measure can be recommended to use for comparisons. First measure is depended on the median value of the series of intervals. For example we denote the interval estimates like

$$\text{Range}_i = \left[\text{Lower Limit}_{(i)} - \text{Upper Limit}_{(i)}\right], i = 1, 2, .., n$$

where n is the number of intervals. Then we calculate the median of the series of ranges and symbolize it as

$$\text{MR} = \text{Median}(\text{Range})$$

The second measure is the ratio of the number of intervals which contain the true value to the total number of intervals. We calculate

$$\text{Ratio} = \frac{k}{n}$$

where k is the number of intervals which contain the true value. It is obvious that the model with the least MR and the greatest Ratio will be the best model. In practice if the true value is equal to the lower or upper limit of the interval estimates this interval it has been assessed that this interval does not contain the true value.

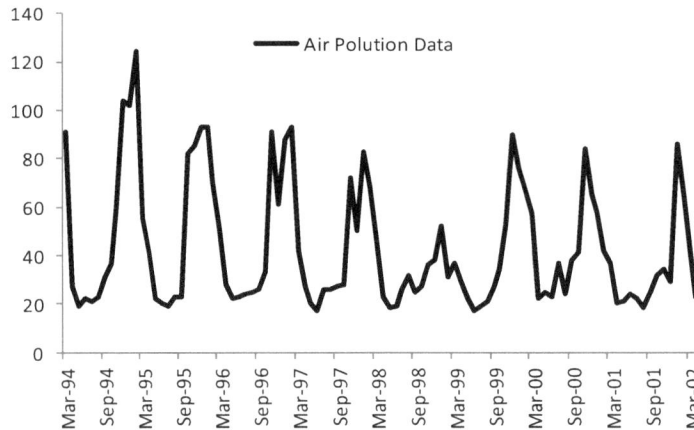

**Figure 1:** The time series plot of the data.

## 5. APPLICATION

The air pollution has been threatening to human health for 50 years. The pollution is caused by the toxic smoke from thermoelectric stations, plants and vehicles. Because of the importance of the issue we want to describe the data of the amounts of carbon dioxide measured in the center and the districts of Ankara between March 1994 and April 2002. We call this time series data as the air-pollution data of Ankara. In Fig. **1**, the time series plot of the data is presented. This graph says that the series changes seasonally and it has also got a trend.

This graph also recommends the seasonal period to be 12. The first and the first seasonal differences of the series are taken for having the stationary. Fig. **2** shows the time series plot of the stationary series.

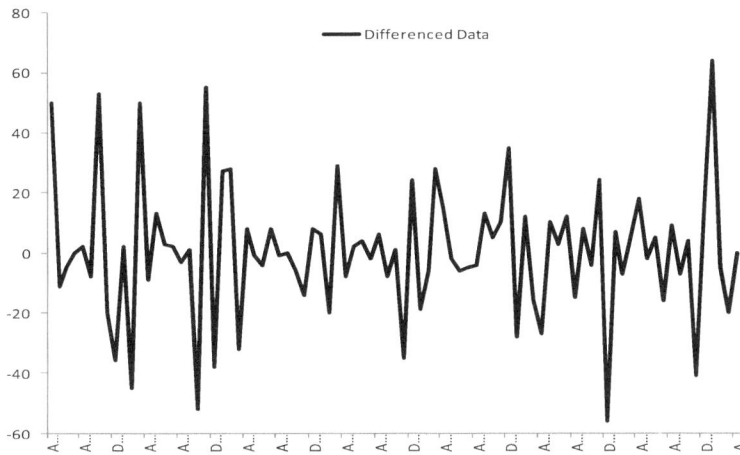

**Figure 2:** The time series plot of the data after taking first and first seasonal differences.

Then we need to determine the degree of SARIMA model. In order to be able to do this, the autocorrelation and the partial autocorrelation function of the stationary series are plotted. Due to the assessments of these plots and the measure of goodness of fit, which is AIC and BIC, the model order is defined as p=1, q=1, Q=1 and P=0. Because of the differences d=1 and D=1 are taken. Then the whole model order is $SARIMA(1,1,1) \times (0,1,1)$. The model equation is

$$(1 - \varphi_1 B)X_t = (1 - \theta_1 B)(1 - \Theta_1 B^{12})a_t$$

The resulting parameter estimations obtained from SPSS application is summarized in Table **1**. All the coefficients are found statistically significant.

**Table 1:** The results of the analysis

| Parameters | Coefficients | s. errors | t-values | p-values |
|---|---|---|---|---|
| AR(1) | 0.316 | 0.111 | 2.843 | 0.006 |
| MA(1) | 0.970 | 0.072 | 13.476 | 0.000 |
| SMA(1) | 0.714 | 0.135 | 5.304 | 0.000 |

In model control stage it has been approved that the residual is a white noise. The predicted and prediction intervals are also calculated from the analysis.

Now we apply the data Fuzzy SARIMA. The residuals and the parameters estimations from the first application are the inputs of the application of Fuzzy SARIMA. These parameter estimations are taken the central values of the fuzzy parameters in Fuzzy SARIMA model. Due to these central values the dispersion measures of the fuzzy parameters is getting by solving the following minimization problem.

$$Min \ S = \sum_{t=1}^{k} F_t$$

*subject to*;

$$\alpha_1 X_{t-1} - \alpha_2 a_{t-1} - \alpha_3 a_{t-12} + \alpha_4 a_{t-13} + (1-h)(c_1|X_{t-1}| + c_2|a_{t-1}| + c_3|a_{t-12}| + c_4|a_{t-13}|) \geq X_t$$

$$\alpha_1 X_{t-1} - \alpha_2 a_{t-1} - \alpha_3 a_{t-12} + \alpha_4 a_{t-13} - (1-h)(c_1|X_{t-1}| + c_2|a_{t-1}| + c_3|a_{t-12}| + c_4|a_{t-13}|) \leq X_t$$

$$c_i \geq 0, i = 1, 2, 3, 4$$

where $\alpha_1 = \emptyset_1, \alpha_2 = \theta_1, \alpha_3 = \Theta_1 \ \alpha_4 = \theta_1 \times \Theta_1$ and $a_t$ is the residual series from SARIMA. Since the product of two triangular fuzzy numbers is also a fuzzy number, that is, $\tilde{\beta}_4 \cong \tilde{\beta}_2 \times \tilde{\beta}_3$ is taken as a separate fuzzy number in the optimization algorithm. The results are;

$$\alpha_1 = 0.316, \alpha_2 = 0.970, \alpha_3 = 0.714, \alpha_4 = 0.970 \times 0.714 = 0.6926$$

For solving the optimization problem a MATLAB program is used. The results of the program give the scale (dispersion) parameters of the coefficients. These are

$$c_1 = 0 \qquad c_2 = 0.4709 \qquad c_3 = 0.2387 \qquad c_4 = 0.3402$$

Finally these central vales and the scale parameters are used to get the interval estimates by using another MATLAB program. The final result of Fuzzy SARIMA model is

$$\widetilde{X_t} = < 0.316, 0 > X_{t-1} - < 0.970, 0.4709 > a_{t-1} - < 0.714, 0.2387 > a_{t-12} + < 0.6926, 0.3402 > a_{t-13} + a_t$$

The statistics MR and Ration are calculated for the two approaches and presented in Table **2**.

**Table 2:** MR and Ratio values

| Model | MR | Ratio |
|---|---|---|
| *SARIMA* | 63.0000 | 0.9459 |
| *FSARIMA* | 20.1021 | 1.0000 |

According to the Table **2**, we conclude one more time that the interval estimates from Fuzzy SARIMA are narrower and include the true value. It means that Fuzzy SARIMA improves the results.

## 6. CONCLUSION

The comparisons of interval estimates are generally being done by graphical methods in the literature. This comparison technique is so subjective that the judgment of different researcher can be different. Get rid of this disadvantages we propose two new criteria which can be used to compare interval estimates. Our application Fuzzy ARIMA and Fuzzy SARIMA have been produced better results than the traditional methods. The aim of using Fuzzy ARIMA and Fuzzy SARIMA is to get interval estimates. When the normality assumption is already satisfied ARIMA and SARIMA also provide interval estimates. According to the results of our analysis it has been concluded that Fuzzy ARIMA and Fuzzy SARIMA improves the interval estimates. These interval estimates are narrower and most of the intervals include the true value and account the direction accuracy. Therefore this assessment is more informative.

# REFERENCES

[1]    L.A. Zadeh, "Fuzzy Sets," *Inform and Control*, vol. 8, pp. 338-353, 1965.
[2]    H. Tanaka, S. Uejima, K. Asai, "Linear regression analysis with fuzzy models", *IEEE Trans. Systems Man, Cybern B*, vol. 12, no. 6, pp. 903-907, 1982.
[3]    F.M. Tseng, G.H. Tzeng, H.C. Yu, and B.J.C. Yuan, "Fuzzy ARIMA model for forecasting the foreign Exchange market," *Fuzzy Set Syst*, vol. 118, pp. 9-19, 2001.
[4]    G.E.P. Box, and G.M. Jenkins, *Time series analysis: Forecasting and control*. San Francisco: Holden-Day, 1970.
[5]    F.M. Tseng, G.H. Tzeng, and H.C. Yu, "Fuzzy seasonal time series for forecasting the production value of the mechanical industry in Taiwan," *Technol Forecast Soc*, vol. 60, pp. 263-273, 1999.

# CHAPTER 9

## The Effect of the Length of Interval in Fuzzy Time Series Models on Forecasting

## Erol Eğrioğlu[1] and Cagdas Hakan Aladag[2,*]

*[1]Ondokuz Mayis University, Faculty of Arts and Science, Department of Statistics, 55139, Samsun, Turkey and [2]Hacettepe University, Faculty of Science, Department of Statistics, 06800, Ankara, Turkey*

**Abstract:** Due to the vagueness that they contain in their observations, fuzzy time series models worked in two main categories such as first order and high order models, has an ever expending field of study. Fuzzy time series analysis method is highly effective in uncovering the relations of this type of time series structure. In the implementation of fuzzy time series methods, it is crucial to determine the model order in terms of forecasting performance. Besides, regardless of the model order, the length of interval determined in the partition phase of the universe of discourse, greatly affects forecasting performance. Therefore, there have been numerous studies focusing on determining the length of interval in the literature. This study aims to introduce the significance of interval length determination in fuzzy time series analysis method on forecasting performance. For this purpose, related methods are introduced, implementation of two real time series is shown and some comparisons between methods are made and finally obtained results are discussed.

**Keywords:** Fuzzy time series, Forecasting, Length of interval, Optimization.

## 1. INTRODUCTION

The implementation of fuzzy time series methods in most common time series with vagueness gives successful results. As well as its effective forecasting performance and not containing constraints found in conventional approaches make fuzzy time series approach applicable and attractive.

Fuzzy time series, which are widely used nowadays, was first proposed by Song and Chissom [1-3]. This followed by simpler methods such as Sullivan and Woodall's models [4] based on Markov model and Chen's model [5] without needing operational matrix processes and using fuzzy logic relation and group relation tables in the relation determination. Additionally, Huarng [6] proposed average and distribution based approaches and he showed that lengths of interval determined in the partition stage of universe of discourse have effect on forecasting performance. Eğrioğlu *et al.* [7] used univariate restricted optimization which minimizes the forecasting error in the determination of the length of interval.

It is a well known fact that model order affects the forecasting performance in fuzzy time methods. Therefore, researchers have used various high order models in the analysis of fuzzy time series. Chen [8], Aladag *et al.* [9] and Eğrioğlu *et al.* [7] aimed to improve forecasting performance using high order models with proposed fuzzy time series analyses approaches. In their studies, while Chen [8] and Eğrioğlu *et al.* [7] determined the fuzzy relations with fuzzy logic relations and group relations tables, Aladag *et al.* [9] used artificial neural networks.

All fuzzy time series methods, whether first order or high order, are sensitive to length of interval determined in the partition stage of universe of discourse in terms of forecasting accuracy. Therefore, interval lengths should be selected in a way that provides the best forecasting accuracy. In the literature, several studies were put forward concerning determination of interval lengths. While the length of interval has been determined subjectively in some approaches, others have tried to improve forecasting performance by bringing more systematic approaches. It should be noted that selection of larger intervals will destroy fluctuation in the series and prevent the emergence of existing relations. Also selection of smaller intervals will eliminate the difference between conventional analysis and fuzzy time series.

***Address correspondence to Cagdas Hakan Aladag:** Hacettepe University, Faculty of Science, Department of Statistics, 06800, Ankara, Turkey; E-mail: chaladag@gmail.com

In the second section of the study, fuzzy time series and its definitions are introduced. Third section introduces first order and high order approaches proposed by Chen [8] and approaches proposed by Huarng [6] and Eğrioğlu *et al.* [7]. Fourth section deals with the application and the application results of the approaches concerning determination of interval length. Finally, the last section provides a conclusion.

## 2. FUZZY TIME SERIES

The definition of fuzzy time series was firstly introduced by Song and Chissom [1, 2]. In fuzzy time series approaches, the validation of theoretical assumptions does not need to be checked just as in conventional time series procedures. The most important advantage of fuzzy time series approaches is to be able to work with a very small set of data and not to require the linearity assumption. Some basic definitions of fuzzy time series can be given as follows [7]:

Let $U$ be the universe of discourse, where $U = \{u_1, u_2, ..., u_b\}$. A fuzzy set $A_i$ of $U$ is defined as $A_i = f_{A_i}(u_1)/u_1 + f_{A_i}(u_2)/u_2 + \cdots + f_{A_i}(u_b)/u_b$, where $f_{A_i}$ is the membership function of the fuzzy set $A_i; f_{A_i} : U \rightarrow [0,1]$. $u_a$ is a generic element of fuzzy set $A_i; f_{A_i}(u_a)$ is the degree of belongingness of $u_a$ to $A_i; f_{A_i}(u_a) \in [0,1]$ and $1 \leq a \leq b$.

**Definition 1.** Fuzzy time series Let $Y(t)(t = ..., 0, 1, 2, ...)$ a subset of real numbers, be the universe of discourse by which fuzzy sets $f_j(t)$ are defined. If $F(t)$ is a collection of $f_1(t), f_2(t), ...$ then $F(t)$ is called a fuzzy time series defined on $Y(t)$.

**Definition 2.** Fuzzy time series relationships assume that $F(t)$ is caused only by $F(t-1)$, then the relationship can be expressed as: $F(t) = F(t-1) * R(t, t-1)$, which is the fuzzy relationship between $F(t)$ and $F(t-1)$, where * represents as an operator. To sum up, let $F(t-1) = A_i$ and $F(t) = A_j$. The fuzzy logical relationship between $F(t)$ and $F(t-1)$ can be denoted as $A_i \rightarrow A_j$ where $A_i$ refers to the left-hand side and $A_j$ refers to the right-hand side of the fuzzy logical relationship. Furthermore, these fuzzy logical relationships can be grouped to establish different fuzzy relationship.

**Definition 3.** Let $F(t)$ be a fuzzy time series. If $F(t)$ is a caused by $F(t-1), F(t-2), ..., F(t-m)$, then this fuzzy logical relationship is represented by

$$F(t-m), ..., F(t-2), F(t-1) \rightarrow F(t),$$

and it is called the $m^{th}$ order fuzzy time series forecasting model.

## 3. THE RELATED METHODS AVAILABLE IN THE LITERATURE

In this section, the methods which were used in the analysis are introduced to emphasize the effect of interval length determination on forecasting performance in fuzzy time series analyses.

### 3.1. Chen's First Order Fuzzy Time Series Method

The method proposed by Chen [5] is based on fuzzy logical relationships and group relation tables. In this method, the lengths of intervals used in partitioning universe of discourse are determined subjectively and the effect of these intervals on forecasting is evident. The following algorithm can be given for the first order fuzzy time series forecasting model proposed by Chen [5]:

**Step 1.** Define the discourse of universe and subintervals. Based on min and max values in the data set, $D_{min}$ and $D_{max}$ variables are defined. Then choose two arbitrary positive numbers which are $D_1$ and $D_2$ in order to divide the interval evenly.

$$U = [D_{min} - D_1, D_{max} + D_2].$$                                                        **(1)**

*Step 2.* Define fuzzy sets based on the universe of discourse and fuzzify the historical data.

*Step 3.* Fuzzify observed rules.

*Step 4.* Establish fuzzy logical relationships and group them based on the current states of the data of the fuzzy logical relationships.

For example, $A_1 \rightarrow A_2, A_1 \rightarrow A_1, A_1 \rightarrow A_3$ , can be grouped as: $A_1 \rightarrow A_2, A_3, A_1$ .

*Step 5.* Forecast.

$$\text{Let } F(t-1) = A_i$$

Case 1: There is only one fuzzy logical relationship in the fuzzy logical relationship sequence. If $A_i \rightarrow A_j$ , then $F(t)$ , forecast value, is equal to $A_j$ .

Case 2 : If $A_i \rightarrow A_i, A_j, \ldots, A_k$ , then $F(t)$ , forecast value, is equal to $A_i, A_j, \ldots, A_k$ .

Case 3: If $A_i \rightarrow empty$ , then $F(t)$ , forecast value, is equal to $A_i$

*Step 6.* Defuzzify.

Apply "Centroid" method to get the results. This procedure (also called center of area, center of gravity) is the most often adopted method of defuzzification. Suppose that the fuzzy forecast of F(t) is $A_k$ . The defuzzyfied forecast is equal to the midpoint of the interval which corresponds to $A_k$ .

### 3.2. Chen's High Order Fuzzy Time Series Method

Chen proposed a method based on high order fuzzy time series which enable to obtain forecasts [8]. The method proposed by Chen [8] produces more accurate forecasts than the first order fuzzy time series methods. In this method, the lengths of intervals are determined subjectively. This affects forecasting accuracy as in the other methods. The model given in definition 3 can be analyzed by the high order fuzzy time series approach. The steps of the algorithm of the method proposed by Chen can be given as follows [8]:

*Step 1.* Define the discourse of universe and subintervals. Based on min and max values in the data set, $D_{min}$ and $D_{max}$ variables are defined. Then choose two arbitrary positive numbers which are $D_1$ and $D_2$ in order to divide the interval evenly.

$$U = \left[ D_{min} - D_1, D_{max} + D_2 \right]. \tag{2}$$

*Step 2.* Define fuzzy sets based on the universe of discourse and fuzzify the historical data.

*Step 3.* Fuzzify observed rules.

*Step 4.* Establish fuzzy logical relationships and group them based on the current states of the data of the fuzzy logical relationships. Based on the linguistically defined variables, $k^{th}$ order fuzzy logical relationship $A_{ik}, A_{i(k-1)}, \ldots, A_{il} \rightarrow A_j$ can be established. For example, the values of the year $i-1$ and $i$ corresponds to fuzzy values $A_a$ and $A_b$ . Also, the values of the year $i+1$ corresponds to fuzzy value $A_j$ . Therefore, $2^{th}$ order fuzzy logical relationship can be written as $A_a, A_b \rightarrow A_j$ . In a similar manner, the more high order fuzzy logical relationships and fuzzy logical groups for 3th, 4th and other high orders are constructed.

*Step 5.* Forecast and Defuzzify.

In here, fuzzy values are defuzzyfied and forecasts are done.

- If the $k^{\text{th}}$ order fuzzified history time series for year $i$ are $A_{ik}, A_{i(k-1)}, ...,$ and $A_{il}$, where $k \geq 2$, and there is the following fuzzy logical relationship in the $k^{\text{th}}$ order fuzzy logical relationship groups shown as follows:

$$A_{ik}, A_{i(k-1)}, ..., A_{il} \rightarrow A_j \qquad (3)$$

where $A_{ik}, A_{i(k-1)}, ..., A_{il}$ and $A_j$, are fuzzy sets, and the maximum membership value of $A_j$ occurs at interval $u_j$, and the midpoint of $u_j$ is $m_j$, then the forecasted time series of year $i$ is $m_j$.

- If the $k$ th order fuzzified history time series for year $i$ are $A_{ik}, A_{i(k-1)}, ...,$ and $A_{il}$, where $k \geq 2$, and there is the following fuzzy logical relationship in the $k$ th order fuzzy logical relationship groups shown as follows:

$$A_{ik}, A_{i(k-1)}, ..., A_{il} \rightarrow A_{j1}$$
$$A_{ik}, A_{i(k-1)}, ..., A_{il} \rightarrow A_{j2}$$
$$\vdots \qquad (4)$$
$$A_{ik}, A_{i(k-1)}, ..., A_{il} \rightarrow A_{jp}$$

where $A_{ik}, A_{i(k-1)}, ..., A_{il}, A_{j1}, A_{j2}, ...,$ and $A_{jp}$, are fuzzy sets, then we can see that there is an ambiguity to forecast the time series of year $i$ (*i.e.* the fuzzy data of year $i$ may be $A_{j1}$ or $A_{j2}$ or... $A_{jp}$). In this case, we must find higher order fuzzified history time series for year $i$, such that there is no ambiguity to forecast the time series of year $i$. Assume that there exists an integer $m$ that can resolve this ambiguity, where $m \geq k$, such that $m$ th order fuzzified time series of year $i$ are $A_{im}, A_{i(m-1)}, ...,$ and $A_{il}$, and there is the following fuzzy logical relationship in the $m$ th order fuzzy logical relationship groups shown as follows:

$$A_{im}, A_{i(m-1)}, ..., A_{il} \rightarrow A_{j1} \qquad (5)$$

where $A_{im}, A_{i(m-1)}, ..., A_{il}$ and $A_j$ are fuzzy sets, and the maximum membership value of $A_j$ occurs at interval $u_j$, and the midpoint of $u_j$ is $m_j$, then the forecasted time series of year $i$ is $m_j$.

- If the $k^{\text{th}}$ order fuzzified history time series for year $i$ are $A_{ik}, A_{i(k-1)}, ...,$ and $A_{il}$, where $k \geq 2$, and there is the following fuzzy logical relationship in the $k^{\text{th}}$ order fuzzy logical relationship groups in which the right hand side of the fuzzy logical relationship is empty shown as follows:

$$A_{ik}, A_{i(k-1)}, ..., A_{il} \rightarrow \# \qquad (6)$$

where $A_{ik}, A_{i(k-1)}, ...,$ and $A_{il}$, are fuzzy sets, and the maximum membership values of $A_{ik}, A_{i(k-1)}, ...,$ and $A_{il}$ occur at intervals $u_{ik}, u_{i(k-1)}, ...,$ and $u_{il}$, respectively, and the midpoint of $u_{ik}, u_{i(k-1)}, ...,$ and $u_{il}$ are $m_{ik}, m_{i(k-1)}, ...,$ and $m_{il}$, respectively then the forecasted time series of year $i$ is calculated as follows:

$$\frac{1 \times m_{ik} + 2 \times m_{i(k-1)} + \cdots + k \times m_{il}}{1 + 2 + \cdots + k} \qquad (7)$$

## 3.3. Huarng's Methods

In fuzzy time series approach, the selection of interval length is a crucial decision. Selection of greater interval length destroys the fluctuation in the time series and smaller interval lengths eliminate the discrimination between conventional analysis and fuzzy time series approach. Therefore, the length of interval should be moderate and use smallest RMSE value. For this reason, Huarng [6] proposed two approaches based on average and distribution. These approaches are explained with the examples given below.

### 3.3.1. Selection of Interval Length with Distribution Based Approach

In distribution based approach, interval length of the initial differences of time series is determined based on distribution. In order to better understand this approach, let time series values be 30, 50, 80, 100 and 70 respectively. First, absolute value of initial differences is calculated. These values are found as 20, 30, 40, 29 and 30 for sample series. Then, the average absolute value of the initial differences is calculated as $\overline{f} = 28$. Base value is chosen from the base mapping table which is shown in Table **1**.

**Table 1:** Base mapping table

| Range | Base |
|---|---|
| 0.1-1 | 0.1 |
| 1.1-10 | 1 |
| 11-100 | 10 |
| 101-1000 | 100 |
| 1001-10000 | 1000 |
| 10001-100000 | 10000 |

As the average absolute value of initial differences is 28 and this value corresponds to 11-100 range in the base table, base value is defined as 10. In this case, as base value is 10, length of interval is chosen including the half of initial difference values. As we have 5 initial differences, the number of differences greater than the selected interval length should be greater than 2.5. The smallest interval length fulfilling the condition will be the interval length of distributed based approach. In our example, the number of initial differences greater than 10 is 3 whereas the number of initial differences greater than 30 is 1, thus the length of interval is defined as 20.

### 3.3.2. Selection of Interval Length with Average Based Approach

Some steps in this method are similar to those in distribution based approach. If the former sample is used in averaged based approach, the absolute values of initial differences and average of these values will be 28 as in the distribution based approach. Then the half of the average is calculated as 28/2=14. As the value of 14 corresponds to 11-100 range in base table, base value is taken as 10 and the value of 14 is round up in accordance with the value of 10. In other words, in this approach, the length of interval is taken as 10 which is a rounded up value of 14. If half of the average was 37, the base value would be 10 and the length of interval would be taken as 40.

### 3.4. The Method Proposed by Eğrioğlu *et al.*

In fuzzy time series approaches, at the forecasting process, the length of intervals affects the forecasting performance. Hence, it is important to choose an effective length of intervals for improving forecasting accuracy in fuzzy time series approaches. For this reason, Eğrioğlu *et al.* [7] proposed an approach based on optimization of interval. This method optimizes the length of interval by following the algorithm of Chen's [8] method. In the optimization process, Eğrioğlu *et al.* [7] used a MATLAB function called "fminbnd" which minimizes MSE. The function "fminbnd" is used to find minimum of a single variable function on a fixed interval. It finds a minimum for a problem specified by

$$\min_{x} f(x) \text{ subject to } x_1 < x < x_2 \qquad (8)$$

$x$, $x_1$, and $x_2$ are scalars and $f(x)$ is a function that returns a scalar. In MATLAB, $\hat{x} = \text{fminbnd}\left(f(x), x_1, x_2\right)$ returns a value $\hat{x}$ that is a local minimum of the scalar valued function $f(x)$ in the interval $x_1 < x < x_2$. In other words, to find the minimum of the function $f(x)$ in the interval $(x_1, x_2)$,

$$a = \text{fminbnd}\,(f(x), x_1, x_2) \tag{9}$$

can be used in MATLAB. $f(a)$ gives the local minimum value in the interval $(x_1, x_2)$.

The algorithm used by "fminbnd" is based on golden section search introduced by Kiefer and parabolic interpolation [10]. Unless the left endpoint $x_1$ is very close to the right endpoint $x_2$, "fminbnd" never evaluates $f(x)$ at the endpoints, so $f(x)$ need only be defined for $x$ in the interval $x_1 < x < x_2$. If the minimum actually occurs at $x_1$ or $x_2$, fminbnd returns an interior point at a distance of no more than $2*TolX$ from $x_1$ or $x_2$, where $TolX$ is the termination tolerance. See [11] or [12] for the details about the algorithm.

## 4. THE IMPLEMENTATION

In this section, in order to get a better view the impact of the length of interval on forecasting performance of fuzzy time series approach, "the number of instructor per faculty at higher education institutions in Turkey between the years 1977 and 2001" and "the amount of jet fuel consumption between January 2005 and December 2007" time series are analyzed. First order models proposed by Chen [5], Huarng [6] and Chen [8], high order models proposed by Aladag *et al.* [9] and Eğrioğlu *et al.* are used in the application.

The results obtained from all these approaches are compared in terms of RMSE values which are widely used in the literature. The equation is given below.

$$RMSE = \sqrt{\frac{\sum_{i=1}^{n}(forecast_t - actual_t)^2}{n}} \tag{10}$$

### 4.1. Application 1

First, time series of "the number of instructor per faculty at higher education institution in Turkey between the years 1977 and 2001" is analyzed. The observation are presented in Table **2**. The graph of this time series is given in Fig. **1**.

Some of the features and parameters of the current analysis are as follows:

- $D_{min} = 50,0429$ and $D_{max} = 84.4829$

- $D_1 = 0.0429$ and $D_2 = 0.5172$

- $U = [D_{min} - D_1, D_{max} + D_2]$, $U = [50, 85]$

- In the methods introduced in [5], [8] and [9], the lengths of interval were taken as 2, 3, 4 and 5.

- The average of absolute initial differences with 24 observations for the average and distribution based approach proposed by Huarng [6] is 3.4995 and half of this value is 1.7498. This value corresponds to 1 in the base table. In this case, the length of interval for both average based and distribution based approaches is found as 2.

- In the method proposed by Aladag *et al.* [9], hidden layer unit numbers in artificial neural network in which fuzzy relations are determined, are limited from 1 to 8.

- In high order models introduced in [7-9], model orders are taken between 1 and 4.

- In the method proposed by Eğrioğlu *et al.* [7] in which the length of intervals are obtained using single variable restricted optimization, the length of interval is optimized using single variable restricted optimization algorithm with MATLAB "fminbnd" function and minimizing RMSE value. In order to show "*la*" integral length, optimization problem can be expressed as follows:

$$\min \ RMSE(la)$$
$$constrained\ to: \ la \in \left[2\ ,\ 5\right] \tag{11}$$

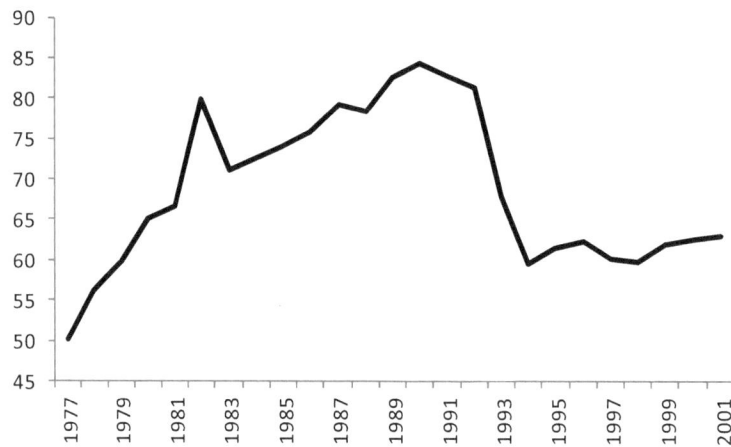

**Figure 1:** The number of instructor per faculty at higher education institutions in Turkey between the years 1977 and 2001.

**Table 2:** The number of instructor and absolute initial differences series

| Years | Instructor | Absolute Difference |
|-------|-----------|---------------------|
| 1977 | 50.0429 | |
| 1978 | 56.0775 | 6.0346 |
| 1979 | 59.6513 | 3.5738 |
| 1980 | 65.1619 | 5.5106 |
| 1981 | 66.5359 | 1.3740 |
| 1982 | 79.9047 | 13.3688 |
| 1983 | 71.0944 | 8.8103 |
| 1984 | 72.6788 | 1.5844 |
| 1985 | 74.0903 | 1.4115 |
| 1986 | 75.7205 | 1.6302 |
| 1987 | 79.2886 | 3.5681 |
| 1988 | 78.4130 | 0.8756 |
| 1989 | 82.7622 | 4.3492 |
| 1990 | 84.4828 | 1.7206 |
| 1991 | 82.8584 | 1.6244 |
| 1992 | 81.3594 | 1.4990 |
| 1993 | 67.9600 | 13.3994 |
| 1994 | 59.4952 | 8.4648 |
| 1995 | 61.5165 | 2.0213 |
| 1996 | 62.3464 | 0.8299 |
| 1997 | 60.1931 | 2.1533 |
| 1998 | 59.7703 | 0.4228 |
| 1999 | 61.9221 | 2.1518 |

*Table 2: cont….*

|  |  |  |
|---|---|---|
| 2000 | 62.6199 | 0.6978 |
| 2001 | 63.0327 | 0.4128 |
|  | *mean* | *3.4996* |
|  | *Half of mean* | *1.7498* |

The results obtained from the analyses in accordance with these features and parameters for the methods given in [5-9] are presented in Table **3-7**, respectively. In Table **6**, first column caption "HLNN" represents number of units in the hidden layer.

**Table 3:** The RMSE values obtained from the method introduced in [5]

| | Length of interval | | |
|---|---|---|---|
| *2* | *3* | *4* | *5* |
| 3.6079 | 4.4814 | 4.0944 | 4.9170 |

**Table 4:** The RMSE values obtained from the method introduced in [8]

| | Length of interval | | | |
|---|---|---|---|---|
| *Orders* | *2* | *3* | *4* | *5* |
| *2* | 0.5877 | 0.9728 | 1.1129 | 1.4783 |
| *3* | 0.5847 | 0.9780 | 1.1355 | 1.4403 |
| *4* | 0.5974 | 0.9575 | 1.1342 | 1.3830 |

**Table 5:** The RMSE values obtained from the method introduced in [6]

| Length of interval $2^{a,b}$ |
|---|
| 3.6079 |

[a] Distribution based approach.

[b] Mean based approach.

**Table 6:** The RMSE values obtained from the method introduced in [9]

| | | Length of interval | | | |
|---|---|---|---|---|---|
| *HLNN* | *Orders* | *2* | *3* | *4* | *5* |
| *1* | *2* | 3.7211 | 4.5633 | 4.0969 | 4.6580 |
| | *3* | 3.6533 | 3.6592 | 3.9552 | 3.8275 |
| | *4* | 3.9699 | 3.8085 | 3.2806 | 3.8743 |
| *2* | *2* | 3.3287 | 3.4207 | 3.2376 | 3.6787 |
| | *3* | 3.5394 | 3.1535 | 3.9552 | 3.8275 |
| | *4* | 3.2186 | 3.3805 | 2.6373 | 3.4145 |
| *3* | *2* | 3.0622 | 3.5978 | 3.0122 | 4.2722 |
| | *3* | 1.7489 | 2.7986 | 2.2796 | 2.9862 |
| | *4* | 1.3959 | 1.0651 | 1.7853 | 3.8289 |
| *4* | *2* | 1.3213 | 3.1949 | 1.8652 | 3.6947 |
| | *3* | 0.9083 | 1.1393 | 1.7124 | 3.7739 |
| | *4* | 1.2568 | 1.4132 | 1.1342 | 1.3830 |
| *5* | *2* | 17.2272 | 1.3539 | 16.3715 | 15.1211 |
| | *3* | 0.8188 | 1.2039 | 1.2903 | 2.5515 |
| | *4* | 0.5974 | 0.9575 | 1.1342 | 1.3830 |

*Table 6: cont....*

| | | | | | |
|---|---|---|---|---|---|
| | 2 | 0.7517 | 3.4810 | 3.4090 | 3.1580 |
| 6 | 3 | 0.8188 | 1.2039 | 1.3951 | 2.6114 |
| | 4 | 0.5974 | 0.9575 | 1.1342 | 1.3830 |
| | 2 | 0.8986 | 1.6872 | 1.5424 | 3.0953 |
| 7 | 3 | 0.8188 | 8.4378 | 1.2903 | 3.6314 |
| | 4 | 0.5974 | 0.9575 | 1.1342 | 1.3830 |
| | 2 | 0.8122 | 1.1117 | 1.5424 | 3.0253 |
| 8 | 3 | 0.8188 | 1.1127 | 1.3951 | 2.6114 |
| | 4 | 0.5974 | 0.9575 | 1.1342 | 1.3830 |

**Table 7:** The RMSE values obtained from the method introduced in [7]

| Orders | Length of interval | RMSE |
|---|---|---|
| 2 | 2.2216 | 0.5385 |
| 3 | 2.2223 | 0.5456 |
| 4 | 2.2201 | **0.5356** |

When the results obtained from all of the methods are evaluated, it is clearly seen that forecasting performance is sensitive to changes in the length of intervals. However, results of methods reveal that appropriate interval length for analyzed fuzzy time series should be 2 or around 2. Performance errors of the best cases obtained from all these methods are summarized in Table **8**. When Table **8** is analyzed, the best forecasting performance is obtained in 2.2201 interval lengths with 0.5356 RMSE value is obtained in the fourth order optimization based model proposed by Eğrioğlu *et al.* [7]. Forecasts related to best case and graphic of actual series are presented in Fig. **2**. According to Fig. **2**, it is obvious that the forecasts calculated using the method proposed by Eğrioğlu *et al.* [7] are very accurate.

**Table 8:** The comparison of the results obtained from all of the methods

| Method | Order | Length of interval | RMSE |
|---|---|---|---|
| Chen's method [5] | 1 | 2 | 3.6079 |
| Huarng's method [6] | 1 | 2 | 3.6079 |
| Chen's method [8] | 3 | 2 | 0.5847 |
| Aladag *et al.*'s method [9] | 2 | 2 | 0.5456 |
| Eğrioğlu *et al.*'s method [7] | 4 | 2.2201 | 0.5356 |

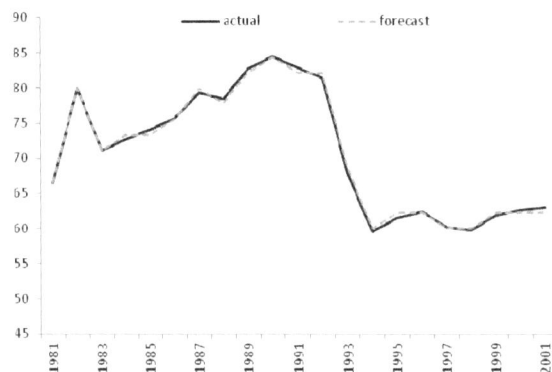

**Figure 2:** Graph of the actual series and forecasts.

## 4.2. Application 2

Secondly, "the amount of jet fuel consumption between January 2005 and December 2007" is analyzed. The observations of the data are presented in Table **9** and their graph is shown in Fig. **3**.

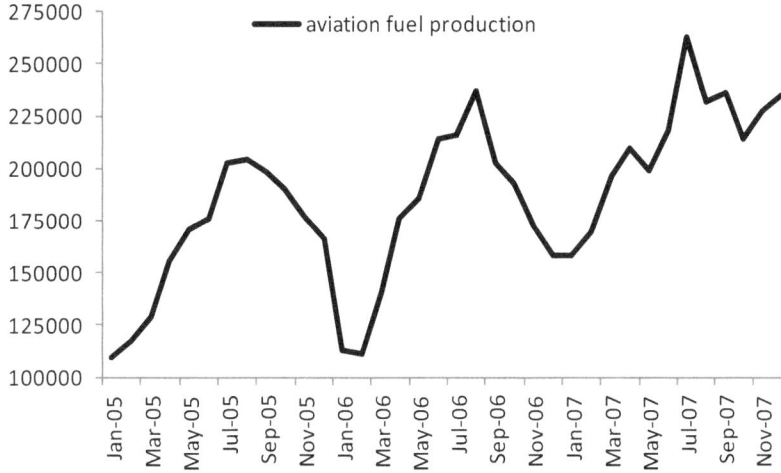

**Figure 3:** The amount of jet fuel consumption between January 2005 and December 2007.

Some of the features and parameters of the current analysis are as follows:

- $D_{min} = 109638$ and $D_{max} = 262617$

- $D_1 = 9638$ and $D_2 = 37383$

- $U = [D_{min} - D_1, D_{max} + D_2]$, $U = [100000, 300000]$

- In [5], [8] and [9] methods, the lengths of interval were taken as 10000, 15000, 20000 and 25000.

- The average of absolute initial differences with 35 observations for the average and distribution based approach proposed by Huarng [6] is 17095.8226 and half of this value is 8547.9143. Base values for these values are 10000 and 1000, respectively. Hence, the length of interval for average based approach is found as 10000 whereas it is found as 9000 in distribution based approach.

- In [9] method, hidden layer unit numbers in artificial neural network in which fuzzy relations are determined, are shifted in 1–8 range.

- In [7-9] high order models, model orders are taken between 1-4.

- In the method proposed by Eğrioğlu *et al.* in which the length of intervals are obtained using single-variable restricted optimization [7], the length of interval is optimized using single-variable restricted optimization algorithm with MATLAB "fminbnd" function and minimizing RMSE value. In order to show "*la*" integral length, optimization problem can be expressed as follows:

min $RMSE(la)$

*constrained to* : $la \in [10000, 25000]$

(12)

**Table 9:** Jet fuel production amount and absolute initial differences series

| Years | Aviation Fuel Production | Absolute Differences |
|---|---|---|
| Jan-05 | 109638 | |
| Feb-05 | 117833 | 8195 |
| Mar-05 | 129249 | 11416 |
| Apr-05 | 155930 | 26681 |
| May-05 | 170577 | 14647 |
| Jun-05 | 176257 | 5680 |
| Jul-05 | 202266 | 26009 |
| Aug-05 | 204692 | 2426 |
| Sep-05 | 198061 | 6631 |
| Oct-05 | 190251 | 7810 |
| Nov-05 | 176401 | 13850 |
| Dec-05 | 165861 | 10540 |
| Jan-06 | 112841 | 53020 |
| Feb-06 | 111585 | 1256 |
| Mar-06 | 140647 | 29062 |
| Apr-06 | 175648 | 35001 |
| May-06 | 185818 | 10170 |
| Jun-06 | 213993 | 28175 |
| Jul-06 | 215639 | 1646 |
| Aug-06 | 237061 | 21422 |
| Sep-06 | 202908 | 34153 |
| Oct-06 | 193121 | 9787 |
| Nov-06 | 172281 | 20840 |
| Dec-06 | 157865 | 14416 |
| Jan-07 | 158228 | 363 |
| Feb-07 | 169590 | 11362 |
| Mar-07 | 196069 | 26479 |
| Apr-07 | 209676 | 13607 |
| May-07 | 199312 | 10364 |
| Jun-07 | 218337 | 19025 |
| Jul-07 | 262617 | 44280 |
| Aug-07 | 231397 | 31220 |
| Sep-07 | 236127 | 4730 |
| Oct-07 | 213711 | 22416 |
| Nov-07 | 227243 | 13532 |
| Dec-07 | 235386 | 8143 |
| | *Mean* | *17095.8286* |
| | *Half of mean* | *8547.9143* |

The results obtained from the analyses in accordance with these features and parameters for [5-9] methods are presented in Table **10-14**, respectively. In Table **13**, first column caption "HLNN" represents number of units in the hidden layer.

**Table 10:** The RMSE values obtained from the method introduced in [5]

| Length of interval | | | |
|---|---|---|---|
| *10000* | *15000* | *20000* | *25000* |
| 15850 | 17467 | 20185 | 19925 |

**Table 11:** The RMSE values obtained from the method introduced in [8]

| | Length of interval | | | |
|---|---|---|---|---|
| *Orders* | *10000* | *15000* | *20000* | *25000* |
| 2 | 2759 | 4685 | 5594 | 6867 |
| 3 | 2701 | 4609 | 5676 | 6820 |
| 4 | 2738 | 4641 | 5668 | 6828 |

**Table 12:** The RMSE values obtained from the method introduced in [6]

| Length of interval | |
|---|---|
| *10000[a]* | *9000[b]* |
| 15850 | 15151 |

[a] Distribution based approach.

[b] Mean based approach.

**Table 13:** The RMSE values obtained from the method introduced in [9]

| | | Length of interval | | | |
|---|---|---|---|---|---|
| *HLNN* | *Orders* | *10000* | *15000* | *20000* | *25000* |
| | 2 | 18755 | 20048 | 20096 | 20088 |
| *1* | 3 | 18020 | 19311 | 19723 | 20340 |
| | 4 | 17998 | 18336 | 19657 | 20623 |
| | 2 | 18492 | 19812 | 19592 | 19449 |
| *2* | 3 | 17465 | 17001 | 17096 | 20272 |
| | 4 | 14927 | 15633 | 16240 | 19429 |
| | 2 | 16047 | 16751 | 19480 | 18732 |
| *3* | 3 | 12222 | 12619 | 12487 | 18499 |
| | 4 | 14927 | 14001 | 13476 | 14869 |
| | 2 | 14842 | 17180 | 18609 | 17307 |
| *4* | 3 | 11445 | 11814 | 12475 | 16932 |
| | 4 | 7631 | 5716 | 9987 | 6906 |
| | 2 | 12251 | 15416 | 19494 | 17791 |
| *5* | 3 | 8375 | 11240 | 10038 | 11561 |
| | 4 | 6359 | 4641 | 5951 | 7345 |
| | 2 | 12025 | 13792 | 17539 | 18537 |
| *6* | 3 | 5193 | 5071 | 8027 | 9913 |
| | 4 | 2738 | 8885 | 7189 | 7345 |
| | 2 | 10636 | 12624 | 17607 | 17797 |
| *7* | 3 | 3363 | 10104 | 12054 | 8829 |
| | 4 | 2738 | 4641 | 5744 | 7345 |
| | 2 | 6851 | 11836 | 17607 | 18067 |
| *8* | 3 | 2701 | 4633 | 5749 | 10580 |
| | 4 | 2738 | 4641 | 5744 | 6906 |

**Table 14:** The RMSE values obtained from the method introduced in [7]

| Orders | Length of interval | RMSE |
|--------|--------------------|------|
| 2 | 10046.38 | 2750 |
| 3 | 10043.06 | 2697 |
| 4 | 10012.65 | 2735 |

As can be seen in the results, determination of interval length plays a crucial role in forecasting. Additionally, results reveal that appropriate interval length for analyzed fuzzy time series should be around 10000. Performance errors of the best cases obtained from all these methods are shown in Table **15**. When Table **15** is analyzed, the best forecasting performance is obtained with 2697 interval length and 10043.46 RMSE value in the third order optimization based model proposed by Eğrioğlu *et al.* [7]. Graph of Forecasts related to the best case and actual series are presented in Fig. **4**. When Fig. **4** is examined, it is clearly seen that the forecasts obtained from the method proposed by Eğrioğlu *et al.* [7] are very accurate.

**Table 15:** The comparison of the results obtained from all of the methods

| Method | Orders | Length of interval | RMSE |
|--------|--------|--------------------|------|
| Chen's method [5] | 1 | 10000 | 15850 |
| Huarng's method [6] | 1 | 9000 | 15151 |
| Chen's method [8] | 3 | 10000 | 2701 |
| Aladag *et al.*'s method [9] | 3 | 10000 | 2701 |
| Eğrioğlu *et al.*'s method [7] | 3 | 10043.06 | 2697 |

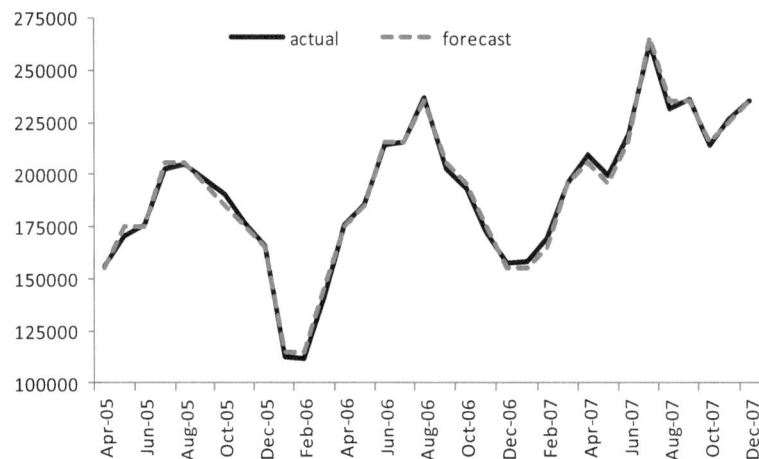

**Figure 4:** Graph of the actual series and forecasts.

## 5. CONCLUSION AND DISCUSSION

The implementation of fuzzy time series in most time series that contain uncertainty gives successful results. As well as its effective forecasting performance and not containing constraints found in traditional approaches make fuzzy time series applicable. In recent years, there have been numerous studies aiming to improve forecasting performance of fuzzy time series analyses.

Many of these studies involve the effect of determination of interval length which constitutes universe of discourse partitioning in the stage of fuzzy time series methods on forecasting performance. All these studies aimed to determine the optimal interval length and thus improve forecasting performance as much as possible.

In this study, some methods proposed for the determination of interval length are introduced, these methods are implemented using two real fuzzy time series, and the effect of interval lengths on forecasting performance is emphasized through the obtained results.

Study results reveal that the lengths of interval play an effective role on forecasting ability of fuzzy time series method. In this context, it is obvious that fuzzy time series will be further discussed in the future.

## REFERENCES

[1]    Q. Song, and B.S. Chissom, "Fuzzy time series and its models," *Fuzzy Set Syst*, vol. 54, pp. 269-277, 1993.

[2]    Q. Song, and B.S.Chissom, "Forecasting enrollments with fuzzy time series - Part I," *Fuzzy Set Syst*, vol. 54, pp. 1-10, 1993.

[3]    Q. Song, and B.S. Chissom, "Forecasting enrollments with fuzzy time series - Part II," *Fuzzy Set Syst,* vol. 62 no. l, pp. 1-8, 1994.

[4]    J. Sullivan, and W.H. Woodall, "A comparison of fuzzy forecasting and Markov modeling," *Fuzzy Set Syst,* vol. 64, no. 3, pp. 279 – 293, 1994.

[5]    S.M. Chen, "Forecasting enrollments based on fuzzy time-series," *Fuzzy Set Syst*, vol. 81, pp. 311-319, 1996.

[6]    K. Huarng, "Effective length of intervals to improve forecasting in fuzzy time series," *Fuzzy Set Syst*, vol. 123, pp. 387-394, 2001.

[7]    E. Eğrioğlu, C.H. Aladag, U. Yolcu, V.R. Uslu, and M.A. Basaran, "Finding an optimal interval length in high order fuzzy time series," *Expert Syst Appl*, vol. 37, pp. 5052-5055, 2010.

[8]    S.M. Chen, "Forecasting enrollments based on high order fuzzy time series," *Cybernet Syst*, vol. 33, pp. 1-16, 2002.

[9]    C.H. Aladag, M.A. Basaran, E. Eğrioğlu, U. Yolcu, and V.R. Uslu, "Forecasting in High Order Fuzzy Times Series by Using Neural Networks to Define Fuzzy Relations," *Expert Syst Appl*, vol. 36, pp. 4228-4231, 2009.

[10]   *J. Kiefer, "Sequential minimax search for a maximum," P Am Math Soc, vol. 4, pp. 502-506, 1953.*

[11]   R.P. Brent, *Algorithms for Minimization without Derivatives*. Prentice-Hall, Englewood Cliff: New Jersey, 1973.

[12]   G.E. Forsythe, M.A. Malcolm, and C.B. Moler, *Computer Methods for Mathematical Computations*. Prentice Hall: New Jersey, 1976.

# CHAPTER 10

# Determining Interval Length in Fuzzy Time Series by Using an Entropy Based Approach

**Cagdas Hakan Aladag**[*], **Irem Degirmenci and Suleyman Gunay**

*Hacettepe University, Faculty of Science, Department of Statistics, 06800, Ankara, Turkey*

**Abstract:** Various theoretical assumptions in conventional time series methods do not need to be checked in fuzzy time series approach. Therefore fuzzy time series are preferred in many applications. The identification of the length of intervals is an important issue and affects the forecasting performance. But in many studies in the literature, the length of intervals is determined randomly. Starting from this point, Huarng [1] has proposed two novel approaches which are based on the distribution and the average to choose a more effective length. Huarng and Yu [2] used a dynamic approach for adjusting lengths of interval. Huarng [3] suggested a different method which is called ratio based lengths of intervals. Cheng *et al.* [4] have proposed a new approach by using entropy. Eğrioğlu *et al.* [5] and Yolcu *et al.* [6] have determined the lengths of intervals by using optimization. At the first stage of the method proposed by Cheng *et al.* [4], a specific method has not been used and classes have been assigned intuitively while classes to which data belong were generating. In this study, the approach proposed by Degirmenci *et al.* [7] is applied to the enrollment data at the University of Alabama and the yearly data of the quantities of clean water used in Istanbul. Then obtained forecasts are compared with those obtained from other methods available in the literature.

**Keywords:** Entropy, Forecasting, Length of interval, Fuzzy c-means clustering, Fuzzy time series.

## 1. INTRODUCTION

Fuzzy set theory was firstly introduced by Zadeh [8] and it has found many application areas since that time. In fuzzy time series, one of the most important application areas is forecasting. Fuzzy time series approach has been introduced as an alternative method for conventional time series models. In contrast to conventional time series methods, various theoretical assumptions do not need to be checked in fuzzy time series approach. The traditional time series approaches require having the linearity assumption and at least 50 observations. The most important advantages of fuzzy time series approach are to be able to work with a few observations and not to require the linearity assumption. In addition to this, the traditional time series methods may be insufficient to forecast the time series whose observations include uncertainties (temperature, stock, *etc.*), even in some cases, traditional approximations can not produce any solution [9]. Fuzzy time series can also be applied to this kind of time series.

Fuzzy time series was firstly introduced by Song and Chissom [10-12]. Afterwards, Sullivan and Woodall [13] proposed an approach based on Markov model. Then, Chen [9] proposed a new method based on Song and Chissom's approach. This method is easier than the method proposed by Song and Chissom and consists of considerably simple calculations. Wang and Lee [14] have revealed a new method which used a different method of fuzzification. In later years, Chen [15] has proposed a new algorithm for high order fuzzy time series forecasting model.

Although many studies have been made about forecasting with fuzzy time series, there are some problems in these studies. The identification of the length of intervals is one of these problems. In fuzzy time series approaches, at the forecasting process, the length of intervals affects the forecasting performance. Hence, it is important to choose an effective length of intervals for improving forecasting accuracy in fuzzy time series approaches. Starting from this point, Huarng [1] has proposed two novel approaches which are based

---

*Address correspondence to Cagdas Hakan Aladag: Hacettepe University, Faculty of Science, Department of Statistics, 06800, Ankara, Turkey; E-mail: chaladag@gmail.com

on the distribution and the average to choose a more effective length. Huarng and Yu [2] used a dynamic approach for adjusting lengths of interval. Huarng [3] suggested a different method which is called ratio based lengths of intervals. Cheng *et al.* [4] have proposed a new approach by using entropy. Eğrioğlu *et al.* [5] and Yolcu *et al.* [6] have determined the lengths of intervals by using optimization.

At the first stage of the method proposed by Cheng *et al.* [4], a specific method has not been used and classes have been assigned intuitively while classes to which data belong were generating. In the approach proposed by Degirmenci *et al.* [7], fuzzy c-means data clustering technique was used to classify data. Afterwards, lengths of intervals were determined by entropy approach.

In this study, the approach proposed by Degirmenci *et al.* [7] is applied to the enrollment data at The University of Alabama which have been used in the most of the other studies in the literature. The yearly data of the quantities of clean water used in Istanbul are also used to reveal that the performance of the proposed approach. In section 2, the fundamental definitions about fuzzy time series are presented. In the third section, Chen's method is given briefly. In subsequent section, the approach proposed by Degirmenci *et al.* [7] and its applications results are presented and compared with other methods in the literature. In final section, the results are discussed.

## 2. FUZZY TIME SERIES

Some basic definitions of fuzzy time series are given as follows [16]:

Let $U$ be the universe of discourse, where $U = \{u_1, u_2, \dots, u_b\}$. A fuzzy set $A_i$ of $U$ is defined as $A_i = f_{A_i}(u_1)/u_1 + f_{A_i}(u_2)/u_2 + \dots + f_{A_i}(u_b)/u_b$, where $f_{A_i}$ is the membership function of the fuzzy set $A_i$ and $f_{A_i}: U \rightarrow [0,1]$. $u_a$ is a generic element of fuzzy set $A_i$; $f_{A_i}(u_a)$ is the degree of belongingness of $u_a$ to $A_i$; $f_{A_i}(u_a) \in [0,1]$ and $1 \leq a \leq b$.

**Definition 1.** Let $Y(t)$ $(t = .,0,1,2,\dots)$ a subset of real numbers, be the universe of discourse by which fuzzy sets $f_j(t)$ are defined. If $F(t)$ is a collection of $f_1(t), f_2(t), \dots$ then $F(t)$ is called a fuzzy time series defined on $Y(t)$.

**Definition 2.** Assume that $F(t)$ is caused only by $F(t-1)$, then the relationship can be expressed as: $F(t) = F(t-1) * R(t, t-1)$, which is the fuzzy relationship $F(t)$ and $F(t-1)$, where * represents as an operator. To sum up, let $F(t-1) = A_i$ and $(t) = A_j$. The fuzzy logical relationship between $F(t)$ and $F(t-1)$ can be denoted as $A_i \rightarrow A_j$ where $A_i$ refers to the left-hand side and $A_j$ refers to the right-hand side of the fuzzy logical relationship. Furthermore, these fuzzy logical relationships can be grouped to establish different fuzzy relationship.

## 3. CHEN'S METHOD [9]

Chen [9] has improved the approximation given by Song and Chissom [10, 11] and he proposes a new method which uses a simpler operation instead of complex matrix operations. The algorithm of Chen's method can be given as follows:

***Step 1.*** Define the universe of discourse and intervals.

Based on the domain issue, The universe of discourse can be defined as: $U = [starting, ending]$. As the length of interval is determined, $U$ can be partitioned into several equally length intervals.

***Step 2.*** Define fuzzy sets based on the universe of discourse and fuzzify the historical data.

***Step 3.*** Fuzzify observed rules.

***Step 4.*** Establish fuzzy logical relationships and group them based on the current states of the data of the fuzzy logical relationships.

For example, $A_1 \rightarrow A_2, A_1 \rightarrow A_1, A_1 \rightarrow A_3$, can be grouped as: $A_1 \rightarrow A_2, A_3, A_1$.

***Step 5.*** Forecast.

Let $F(t-1) = A_i$.

Case 1: There is only one fuzzy logical relationship in the fuzzy logical relationship sequence.

If $A_i \rightarrow A_j$, then $F(t)$, forecast value, is equal to $A_j$.

Case 2: If $A_i \rightarrow A_i, A \quad, ..., A_k$, then $F(t)$, forecast value, is equal to $A_i, A_j, ..., A_k$.

Case 3: If $A_i \rightarrow empty$, then $F(t)$, forecast value, is equal to $A_i$.

***Step 6.*** Defuzzify.

Apply "Centroid" method to get the results. This procedure is the most often adopted method of defuzzification.

## 4. THE METHOD PROPOSED BY DEGIRMENCI *et al.* [7]

One of the most important problems in the fuzzy time series forecasting is determining the lengths of intervals. The length of intervals affects the performance of forecasting significantly. One of the methods for choosing the length of intervals is the entropy approach which was proposed by Cheng *et al.* [4].

The entropy of a probability distribution is a measure of the uncertainty of the distribution [4]. Let $X$ be a random variable. $(p_1, p_2, ..., p_n)$ to the probability distribution of this variable, entropy $H$ is,

$$H = -\sum_{i=1}^{n} p_i ln p_i$$

In this study, Minimize Entropy Principle Approach is used. This approach is preferred method to use when the decision maker does not have any prior knowledge [4].

For partitioning the universal set to the subintervals, determining the threshold between classes of data is needed. In the method proposed by Cheng *et al.* [4], every threshold is obtained by minimizing the entropy. After obtaining the first threshold, the segmentation process is started. The primary threshold firstly divides the data into two classes. At the end of this repeated partitioning with threshold value calculations, the data set is partitioned into any number of fuzzy sets.

Assume that a threshold value is sought for a sample in the range between $x_1$ and $x_2$. An entropy equation with each value of $x$ is written for the regions $[x_1, x]$ and $[x, x_2]$ and the first region is denoted by $p$, the second region is denoted by $q$. An entropy with each value of $x$ in the region between $x_1$ and $x_2$ is expressed as

$$S(x) = p(x)S_p(x) + q(x)S_q(x)$$

where

$$S_p(x) = -[p_1(x)lnp_1(x) + p_2(x)lnp_2(x)]$$

$$S_q(x) = -[q_1(x)lnq_1(x) + q_2(x)lnq_2(x)]$$

where $p_k(x)$ and $q_k(x)$ = conditional probabilities that the class $k$ sample is in the region $[x_1, x]$ and $[x, x_2]$, respectively $p(x)$ and $q(x)$ = probabilities that all samples are in the region $[x_1, x]$ and $[x, x_2]$, respectively. Also $p(x) + q(x) = 1$.

The value of $x$ which has minimum entropy is the optimum threshold value. Specified probability values are calculated as follows:

$$p_k(x) = \frac{n_k(x) + 1}{n(x) + 1}$$

$$q_k(x) = \frac{N_k(x) + 1}{N(x) + 1}$$

$$p(x) = n(x)/n$$

$$q(x) = 1 - p(x)$$

where

$n_k(x)$ = number of class $k$ samples located in $[x_1, x]$

$n(x)$ = the total number of samples located in $[x_1, x]$

$N_k(x)$ = number of class $k$ samples located in $[x, x_2]$

$N(x)$ = the total number of samples located in $[x, x_2]$

$n$ = total number of samples in $[x_1, x_2]$.

While moving $x$ in the region $[x_1, x_2]$, the values of entropy are calculated for each position of $x$. The $x$ value which has minimum entropy in this region is called the primary threshold (PRI) value. By repeating this process, secondary threshold values (SEC1 and SEC2) are determined. Since seven partitions are wanted, tertiary threshold values (TER1, TER2, TER3, and TER4) are found by repeating this process once more. Method's working is shown in Fig. **1**.

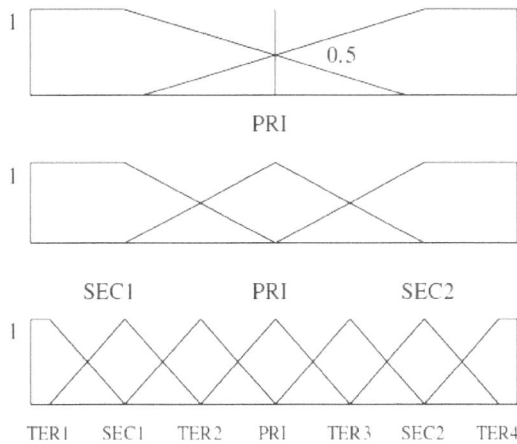

**Figure 1:** The selection of threshold values.

The approach proposed by Degirmenci *et al.* [7]'s working is explained below step by step through Alabama data.

***Step 1.*** Firstly, each data is assigned to a class. In the method proposed by Cheng *et al.* [4], any particular method has not been used about number of classes or classes which each of the data take place in. Instead of this, classes have been determined as intuitively. In the approach proposed by Degirmenci *et al.* [7], fuzzy c-means data clustering technique was used to specify the classes. Since number of class was selected as 3 in the Cheng's method, class number was also identified as 3 in this study. The classes which were obtained as a result of fuzzy c-mean method are given in Table **1**.

**Table 1:** Classes assigned of each year

| Year | Enrollment | Class | Year | Enrollment | Class |
|------|-----------|-------|------|-----------|-------|
| 1971 | 13055 | 1 | 1982 | 15433 | 2 |
| 1972 | 13563 | 1 | 1983 | 15497 | 2 |
| 1973 | 13867 | 1 | 1984 | 15145 | 2 |
| 1974 | 14696 | 2 | 1985 | 15163 | 2 |
| 1975 | 15460 | 2 | 1986 | 15984 | 2 |
| 1976 | 15311 | 2 | 1987 | 16859 | 2 |
| 1977 | 15603 | 2 | 1988 | 18150 | 3 |
| 1978 | 15861 | 2 | 1989 | 18970 | 3 |
| 1979 | 16807 | 2 | 1990 | 19328 | 3 |
| 1980 | 16919 | 2 | 1991 | 19337 | 3 |
| 1981 | 16388 | 2 | 1992 | 18876 | 3 |

***Step 2.*** Threshold values are calculated (PRI, SEC1, SEC2, TER1, TER2, TER3, and TER4).

The entropy of each data is found by entropy equation which was given by Christensen [17] described above. Firstly, the data set must be sorted according to enrollments of each year. Afterwards, entropy values between every two adjacent data are calculated and data which has minimum entropy value is determined. Table **2** shows that when $x = 17535$, the entropy value $S(x)$ is the smallest. So PRI primary threshold value is 17535. In table 2, the values for two random data are given together with the value which has minimum entropy for making this step clearer. By repeating this step, all threshold values are obtained as shown in Table **3**.

As shown in Fig. **1**, seven threshold values are obtained at the end of the three stages in this step. New threshold values can be obtained by repeating this stage. Since seven partitions have been used to forecast in the literature, the number of threshold values are determined as seven.

**Table 2:** Entropy value calculation

| $x$ | (15145+15163)/2=15154 | (16919+18150)/2=17535 | (18876+18970)/2=18923 |
|-----|----------------------|----------------------|----------------------|
| $p_1(x)$ | (3+1)/(5+1)=0.6666 | (3+1)/(17+1)=0.2222 | (3+1)/(19+1)=0.2000 |
| $p_2(x)$ | (2+1)/(5+1)=0.5000 | (14+1)/(17+1)=0.8333 | (14+1)/(19+1)=0.7500 |
| $p_3(x)$ | (0+1)/(5+1)=0.1666 | (0+1)/(17+1)=0.0556 | (2+1)/(19+1)=0.1500 |
| $q_1(x)$ | (0+1)/(17+1)=0.0556 | (0+1)/(5+1)=0.1667 | (0+1)/(3+1)=0.2500 |
| $q_2(x)$ | (12+1)/(17+1)=0.7222 | (0+1)/(5+1)=0.1667 | (0+1)/(3+1)=0.2500 |
| $q_3(x)$ | (5+1)/(17+1)=0.3333 | (5+1)/(5+1)=1 | (3+1)/(3+1)=1 |
| $p(x)$ | 5/22=0.2273 | 17/22=0.7727 | 19/22=0.8636 |
| $q(x)$ | 17/22=0.7727 | 5/22=0.2273 | 3/22=0.1364 |
| $S_p(x)$ | 0.9155 | 0.6468 | 0.8222 |
| $S_q(x)$ | 0.7618 | 0.5973 | 0.6931 |
| $S(x)$ | 0.7967 | 0.6355 | 0.8046 |

**Table 3:** Threshold values

| TER1 | SEC1 | TER2 | PRI | TER3 | SEC2 | TER4 |
|------|------|------|-----|------|------|------|
| 13715 | 14282 | 16889 | 17535 | 18513 | 18923 | 19332 |

***Step 3.*** The lengths of intervals are determined depending on the threshold values in step 2 and the membership function is established. The threshold values are defined as the midpoint of the triangular fuzzy number. With the help of this midpoint, the lower-upper limits of the fuzzy number are determined. The membership function of Minimize Entropy Principle Approach is shown in Table **4**.

**Table 4.** The membership function

| Linguistic value | Lower Bound | Midpoint | Upper Bound | Length of interval |
|------------------|-------------|----------|-------------|-------------------|
| $A_1$ (very very few) | 11280 | 13715 | 14282 | 3002 |
| $A_2$ (ver few) | 13715 | 14282 | 16889 | 3174 |
| $A_3$ (few) | 14282 | 16889 | 17535 | 3253 |
| $A_4$ (moderate) | 16889 | 17535 | 18513 | 1624 |
| $A_5$ (many) | 17535 | 18513 | 18923 | 1388 |
| $A_6$ (very many) | 18513 | 18923 | 19332 | 819 |
| $A_7$ (very very many) | 18923 | 19332 | 21112 | 2189 |

***Step 4.*** The historical data are fuzzified.

The membership degree of each data was calculated to determine its linguistic value according to the membership function in Step 3 and is shown in Table **5**.

**Table 5:** Fuzzified enrollments

| Year | Enrollments | Linguistic value | Year | Enrollments | Linguistic value |
|------|-------------|------------------|------|-------------|------------------|
| 1971 | 13055 | A1 | 1982 | 15433 | A2 |
| 1972 | 13563 | A1 | 1983 | 15497 | A2 |
| 1973 | 13867 | A1 | 1984 | 15145 | A2 |
| 1974 | 14696 | A2 | 1985 | 15163 | A2 |
| 1975 | 15460 | A2 | 1986 | 15984 | A3 |
| 1976 | 15311 | A2 | 1987 | 16859 | A3 |
| 1977 | 15603 | A3 | 1988 | 18150 | A5 |
| 1978 | 15861 | A3 | 1989 | 18970 | A6 |
| 1979 | 16807 | A3 | 1990 | 19328 | A7 |
| 1980 | 16919 | A3 | 1991 | 19337 | A7 |
| 1981 | 16388 | A3 | 1992 | 18876 | A6 |

***Step 5.*** The fuzzy relations are established and the forecasts are obtained. In this step, Chen's method which is introduced in section 2 is applied in the same way.

The results of the approach proposed by Degirmenci *et al.* [7] and other studies in the literature are shown in Table **6**. The last row of the table shows the calculated mean square (MSE) error values. The mean square error (MSE) value used for comparison is calculated as:

$$MSE = \frac{\sum_{i=1}^{n}(Actual\ enrollment_i - Forecasted\ enrollment_i)^2}{n}$$

The approach proposed by Degirmenci *et al.* [7] is compared with other methods under the same conditions. Firstly, the results have been obtained on the same data set that is the annual enrollments of the Alabama University. Also, seven linguistic values have been used for each method. All of them are first order and time invariant models.

**Table 6:** The forecasting results

| Year | Actual enrollments | Song and Chissom [10, 11] | Sullivan and Woodall [13] | Chen [9] | Wang and Lee [14] (w=2) | Cheng *et al.* [4] | Degirmenci *et al.* [7] |
|---|---|---|---|---|---|---|---|
| 1971 | 13055 | | | | | | |
| 1972 | 13563 | 14000 | 13500 | 14000 | | 15430 | 14042 |
| 1973 | 13867 | 14000 | 14500 | 14000 | 13860 | 15430 | 14042 |
| 1974 | 14696 | 14000 | 14500 | 14000 | 13964 | 15430 | 14042 |
| 1975 | 15460 | 15500 | 15231 | 15500 | 14710 | 15430 | 15605 |
| 1976 | 15311 | 16000 | 15563 | 16000 | 15452 | 15430 | 15605 |
| 1977 | 15603 | 16000 | 15563 | 16000 | 15311 | 15430 | 15605 |
| 1978 | 15861 | 16000 | 15500 | 16000 | 15603 | 15430 | 16480 |
| 1979 | 16807 | 16000 | 15500 | 16000 | 15861 | 16889 | 16480 |
| 1980 | 16919 | 16813 | 16684 | 16833 | 16830 | 16871 | 16480 |
| 1981 | 16388 | 16813 | 16684 | 16833 | 16919 | 16871 | 16480 |
| 1982 | 15433 | 16789 | 15500 | 16833 | 16388 | 15447 | 16480 |
| 1983 | 15497 | 16000 | 15563 | 16000 | 15417 | 15430 | 15605 |
| 1984 | 15145 | 16000 | 15563 | 16000 | 15497 | 15430 | 15605 |
| 1985 | 15163 | 16000 | 15563 | 16000 | 15145 | 15430 | 15605 |
| 1986 | 15984 | 16000 | 15563 | 16000 | 15163 | 15430 | 15605 |
| 1987 | 16859 | 16000 | 15500 | 16000 | 15984 | 16889 | 16480 |
| 1988 | 18150 | 16813 | 16577 | 16833 | 16862 | 16871 | 16480 |
| 1989 | 18970 | 19000 | 19500 | 19000 | 18122 | 19333 | 18923 |
| 1990 | 19328 | 19000 | 19500 | 19000 | 18970 | 19333 | 20018 |
| 1991 | 19337 | 19000 | 19500 | 19000 | 19091 | 19333 | 19470 |
| 1992 | 18876 | | | 19000 | 19101 | 19333 | 19470 |
| **MSE** | | 423027 | 386055 | 407507 | 377728 | 446762 | 329839 |

As shown in Table **6**, the method which has minimum MSE value is the approach proposed by Degirmenci *et al.* [7]. In other words, it is seen that this approach gives better results compared to other methods in the literature. The graph of the forecasting results with the real values is shown in Fig. **2**.

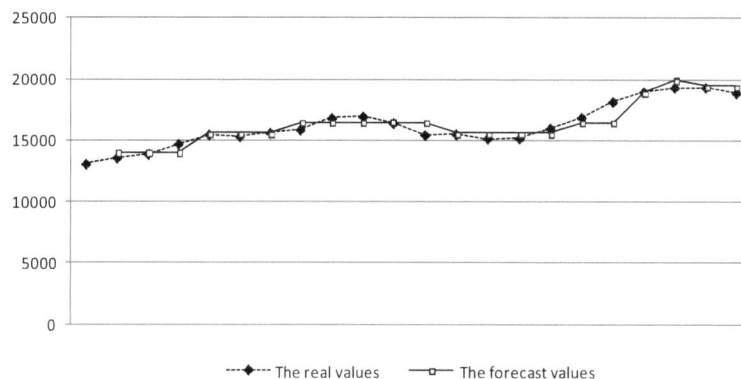

**Figure 2:** The graph of the forecast values and the real values.

The approach proposed by Degirmenci *et al.* [7] is also applied to the yearly data of the quantities of clean water used in Istanbul. Firstly, each data is assigned to a class as shown in Table **7**.

**Table 7:** Classes assigned of each year

| Year | The amount of clean water (1000 m³) | Class |
|------|------|------|
| 1994 | 392 | 2 |
| 1995 | 525 | 1 |
| 1996 | 552 | 1 |
| 1997 | 570 | 1 |
| 1998 | 599 | 1 |
| 1999 | 618 | 1 |
| 2000 | 615 | 1 |
| 2001 | 639 | 1 |
| 2002 | 669 | 3 |
| 2003 | 692 | 3 |
| 2004 | 709 | 3 |
| 2005 | 698 | 3 |
| 2006 | 732 | 3 |
| 2007 | 714 | 3 |
| 2008 | 718 | 3 |

Then the threshold values can be calculated by the Step 2, as shown in Table **8**.

**Table 8:** Threshold values

| TER1 | SEC1 | TER2 | PRI | TER3 | SEC2 | TER4 |
|------|------|------|------|------|------|------|
| 458.5 | 538.5 | 628.5 | 654 | 703.5 | 711.5 | 725 |

Membership function can be established as shown in Table **9**.

**Table 9:** The membership function

| Linguistic value | Lower Bound | Midpoint | Upper Bound | Length of interval |
|------|------|------|------|------|
| $A_1$ (very very few) | 299 | 458.5 | 538.5 | 239.5 |
| $A_2$ (ver few) | 458.5 | 538.5 | 628.5 | 170 |
| $A_3$ (few) | 538.5 | 628.5 | 654 | 115.5 |
| $A_4$ (moderate) | 628.5 | 654 | 703.5 | 75 |
| $A_5$ (many) | 654 | 703.5 | 711.5 | 57.5 |
| $A_6$ (very many) | 703.5 | 711.5 | 725 | 21.5 |
| $A_7$ (very very many) | 711.5 | 725 | 825 | 113.5 |

Then the data can be fuzzified to linguistic values as shown in Table **10**.

**Table 10:** Fuzzified yearly data of the quantities of clean water used in Istanbul

| Year | The amount of clean water (1000 m³) | Linguistic value |
|------|------|------|
| 1994 | 392 | A1 |
| 1995 | 525 | A2 |

*Table 10: cont….*

| 1996 | 552 | A2 |
|------|-----|-----|
| 1997 | 570 | A2 |
| 1998 | 599 | A3 |
| 1999 | 618 | A3 |
| 2000 | 615 | A3 |
| 2001 | 639 | A4 |
| 2002 | 669 | A4 |
| 2003 | 692 | A5 |
| 2004 | 709 | A6 |
| 2005 | 698 | A5 |
| 2006 | 732 | A7 |
| 2007 | 714 | A6 |
| 2008 | 718 | A6 |

Finally, the fuzzy relations are established and the forecasts are calculated according to Step 5. The results are shown in Table **11**.

**Table 11:** The forecasting results

| Year | Actual quantities | Song and Chissom [10, 11] | Chen [9] | Degirmenci *et al.* [7] |
|------|-------------------|---------------------------|----------|-------------------------|
| 1994 | 392 | - | - | - |
| 1995 | 525 | 557 | 529.5 | 557,00 |
| 1996 | 552 | 557 | 557 | 576,63 |
| 1997 | 570 | 612 | 557 | 576,63 |
| 1998 | 599 | 612 | 612 | 576,63 |
| 1999 | 618 | 667 | 612 | 631,13 |
| 2000 | 615 | 667 | 667 | 631,13 |
| 2001 | 639 | 667 | 667 | 631,13 |
| 2002 | 669 | 722 | 667 | 674,38 |
| 2003 | 692 | 722 | 722 | 674,38 |
| 2004 | 709 | 722 | 722 | 741,25 |
| 2005 | 698 | 722 | 722 | 698,50 |
| 2006 | 732 | 694.5 | 722 | 741,25 |
| 2007 | 714 | 722 | 694.5 | 714,25 |
| 2008 | 718 | 722 | 722 | 698,50 |
| *MSE* | | 3877 | 432.32 | 322.50 |

The yearly data of the quantities of clean water used in Istanbul are also used to show the performance of the approach proposed by Degirmenci *et al.* [7]. As seen in Table **11**, the approach proposed by Degirmenci *et al.* [7] gives better results compared to other methods in the literature.

## 5. CONCLUSION

Determination of length of intervals and linguistic values is one of the most important issues encountered in fuzzy time series approach. In order to overcome this problem, various methods have been proposed in the literature. In these studies, subjective approach is usually used to determine the length of intervals. One of these studies is fuzzy time series forecasting proposed by Cheng *et al.* [4]. However, in the method proposed by Cheng *et al.* [4], a specific method was not used during determination of the classes and random selection was made. In this study, the approach proposed by Degirmenci *et al.* [7] is used in which entropy approach and fuzzy c-means data clustering methods are employed to obtain more accurate forecasts. Also in Cheng *et al.*'s method [4], in determining the fuzzy sets of data to which they belong, the data in the intersection of two sets were directly involved in the next cluster. In the method proposed by Degirmenci *et al.* [7], the data in the intersection are involved in the fuzzy set whose membership degree is greater. In this study, this method is applied to the enrollment data at University of Alabama and the yearly data of the quantities of clean water used in Istanbul. It is seen that the method proposed by Degirmenci *et al.* [7] gives better results than some other methods available in the literature. With the used approach, fuzzy time series forecasting method based on entropy has been made more systematic and also more accurate forecasts were obtained.

## REFERENCES

[1]    K. Huarng, "Effective length of intervals to improve forecasting in fuzzy time series," *Fuzzy Set Syst*, vol. 123, pp. 387-394, 2001.

[2]    K. Huarng, and H. Yu, "A dynamic approach to adjusting lengths of intervals in fuzzy time series forecasting," *Intell Data Anal*, vol. 8, no. 1, pp. 3-27, 2004.

[3]    K. Huarng, "Ratio-based lengths of intervals to improve fuzzy time series forecasting," *IEEE Trans Syst Man, Cybern B, Cybern*, vol. 36, pp. 328-340, 2006.

[4]    C.H. Cheng, J.R. Chang, and C.A. Yeh, "Entropy-based and trapezoid fuzzification-based fuzzy time series approaches for forecasting IT Project cost," *Technol Forecast Soc*, vol. 73, pp. 524-542, 2006.

[5]    E. Eğrioğlu, C.H. Aladag, M.A. Basaran, U. Yolcu, and V.R. Uslu, "A new approach based on the optimization of the length of intervals in fuzzy time series," *J Intell Fuzzy Syst*, vol. 22, pp. 15–19, 2011.

[6]    U. Yolcu, E. Eğrioğlu, V.R. Uslu, M.A. Basaran, and C.H. Aladag, "A new approach for determining the length of intervals for fuzzy time series," *Appl Soft Comput*, vol. 9, pp. 647-651, 2009.

[7]    I. Degirmenci, C.H. Aladag, S. Gunay, and E. Eğrioğlu, "Using entropy approach in fuzzy time series forecasting," *Optim Lett*, under review.

[8]    L.A. Zadeh, "Fuzzy Sets," *Inform. and Control*, vol. 8, pp. 338-353, 1965.

[9]    S.M. Chen, "Forecasting enrollments based on fuzzy time-series," *Fuzzy Set Syst*, vol. 81, pp. 311-319, 1996.

[10]   Q. Song, and B.S. Chissom, "Fuzzy time series and its models," *Fuzzy Set Syst*, vol. 54, pp. 269-277, 1993a.

[11]   Q. Song, and B.S. Chissom, "Forecasting enrollments with fuzzy time series - Part I," *Fuzzy Set Syst*, vol. 54, pp. 1-10, 1993b.

[12]   Q. Song, and B.S. Chissom, "Forecasting enrollments with fuzzy time series - Part II," *Fuzzy Set Syst*, vol. 62, pp. 1-8, 1994.

[13]   J. Sullivan, and W.H. Woodall, "A comparison of fuzzy forecasting and Markov modeling," *Fuzzy Set Syst*, vol. 64, pp. 279-293, 1994.

[14]   H.F. Wang, and C.T. Lee, "A method for fuzzy time series analysis-an example for telecommunication demands," In: *IFORS'96*, Vancouver: Canada, July, 8-12, 1996.

[15]   S.M. Chen, "Forecasting enrollments based on high order fuzzy time series," *Cybernet Syst*, vol. 33, pp. 1-16, 2002.

[16]   C.H. Aladag, M.A. Basaran, E. Eğrioğlu, U. Yolcu, and V.R. Uslu, "Forecasting in high order fuzzy times series by using neural networks to define fuzzy relations," *Expert Syst Appl*, vol. 36, no. 3, pp. 4228–4231, 2009.

[17]   R. Christensen, *Entropy Minimax Sourcebook*, v. 1-4. Entropy Ltd.: Lincoln, MA, 1980.

# CHAPTER 11

# An Architecture Selection Method Based on Tabu Search

## Cagdas Hakan Aladag[*]

*Hacettepe University, Faculty of Science, Department of Statistics, 06800, Ankara, Turkey*

**Abstract:** In recent years, the most preferred forecasting method in time series forecasting has been artificial neural networks. In many applications, artificial neural networks have been successfully employed to obtain accurate forecasts in the literature. This approach has been preferred to conventional time series forecasting models because of its easy usage and providing accurate results. On the other hand, there are still some problems with using this method. Fining a good artificial neural network architecture which gives the most accurate forecasts is an important issue when the method is used for forecasting. Although, there are some systematical methods proposed to determine the best architecture, the most preferred method is trial and error method [1]. To solve the architecture selection problem, Aladag [2] also proposed an approach based on tabu search algorithm. In this study, the air pollution in Ankara time series is forecasted by utilizing artificial neural networks and the architecture selection algorithm proposed by Aladag [2] is used to determine the best architecture. The obtained results show that high accuracy level is reached when Aladag's [2] algorithm is employed.

**Keywords:** Architecture selection, Artificial neural networks, Forecasting, Tabu search, Time series.

## 1. INTRODUCTION

Forecasting in time series is an important issue on which many researchers from different disciplines have still working. In the literature, various approaches have been proposed to obtain more accurate forecasts. In recent years, artificial neural networks approach has the most preferred method for forecasting since the method has proved its success in many forecasting applications [3]. However, there are some drawbacks in using artificial neural networks to forecast time series. When this method is utilized for forecasting, selection of the components of the method is a vital issue to reach good forecasting results [4]. Determining the best artificial neural network model can be defined as selection of the components such as architecture structure, learning algorithm and activation function [5]. Selection of the best model, especially determining the best architecture and weights, remains a problem in artificial neural networks applications [6].

When the best model is being searched, an important decision is the selection of architecture structure consists of determining the numbers of neurons in the layers of a network. Various techniques have been proposed to determine the best artificial neural network architecture [2]. Some of them are constructive and pruning algorithm [7], polynomial time algorithm [8], network information criterion [9], iterative construction algorithm [10], a method based on Box-Jenkins analysis [11], a method based on information entropy [12], genetic algorithms [13], the principle component analysis [14], weighted information criterion [5], a deletion/substitution/addition algorithm [15], an architecture selection strategy for autoregressive seasonal time series [16], and design of experiments [17]. In spite of these proposed method, the most preferred method is trial and error [1, 2]. Trial and error approach, however, is not rigorous and offers no guarantee of arriving at a truly optimal structure [6].

Both theoretical and empirical findings in the literature show that combining different methods can be an efficient way to improve forecasts [4]. Therefore, in this study, a hybrid forecasting method combines artificial neural networks and tabu search algorithm is used to forecast the time series of air pollution of Ankara which is the capitol city of Turkey. The time series is forecasted by using artificial neural networks and tabu search algorithm proposed by Aladag [2] is used to determine the best architecture. For

---

*Address correspondence to Cagdas Hakan Aladag:** Hacettepe University, Faculty of Science, Department of Statistics, 06800, Ankara, Turkey; E-mail: chaladag@gmail.com

comparison, the time series is also forecasted using trial and error method which is the most preferred technique in the literature. Then, the obtained results are examined and discussed.

The next section briefly gives artificial neural networks. Section 3 introduces the hybrid approach in which an architecture selection method based tabu search algorithm is employed. Section 4 presents the implementation. Finally, the last section concludes the chapter.

## 2. ARTIFICIAL NEURAL NETWORKS

Artificial neural networks have attracted more and more attention from both academic researcher and industrial practitioners in the recent years [19]. This approach has been widely used to model time series in various fields of applications [20]. Due to ability of modeling both linear and non-linear structures, artificial neural networks have been used as a good alternative method for both linear and non-linear time series forecasting. Zhang *et al.* [21] presented a review of the current status in applications of neural networks for forecasting.

In time series forecasting, artificial neural networks are mathematical models that imitate biological neural networks. Artificial neural networks consist of some elements. Determining the elements of the artificial neural networks issue that affect the forecasting performance of artificial neural networks should be considered carefully [16]. Elements of the artificial neural networks can be given as network architecture, learning algorithm and activation function [22].

For forecasting problems, feed forward neural networks have been widely preferred as architecture structure because of their simple usage and success. The structure of multilayer feed forward neural network is basically given in Fig. **1**. A multilayer feed forward neural network as illustrated in the figure consists of three parts such as input, hidden, and output layers. Each layer consists of neurons. The architecture structure is determined based on deciding the number of neuron in each layer [6].

When artificial neural networks are used for forecasting, one critical decision is to determine the appropriate architecture, that is, the number of layers, number of nodes in each layer [2]. However, in the literature, there are not general rules for determining the best architecture [6]. Although there are some systematic approaches in the literature, the most preferred method to find the best architecture is trial and error method.

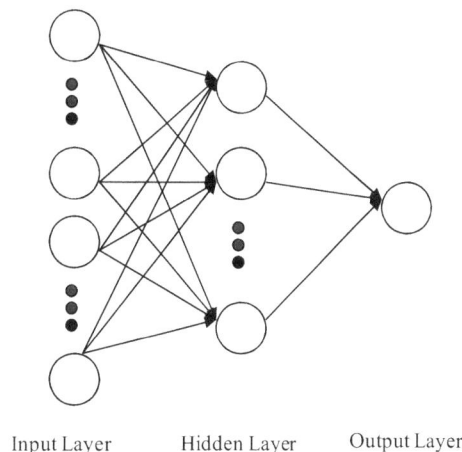

Input Layer          Hidden Layer          Output Layer

**Figure 1:** Multilayer feed forward artificial neural network with one output neuron.

Learning of an artificial neural network for a specific task is equivalent to finding the values of all weights such that the desired output is generated by the corresponding input. Various training algorithms have been used for the determination of the optimal weights values. One of the training algorithms is Levenberg Marquardt algorithm [23]. This algorithm is used as training algorithm in most optimization packages since it is an efficient nonlinear optimization method.

Another element of the artificial neural networks is the activation function. It determines the relationship between inputs and outputs of a network. In general, the activation function introduces a degree of the non-linearity that is valuable in most of ANN applications. The well known activation functions are logistic, hyperbolic tangent, sine (or cosine) and the linear functions. Among them, logistic activation function is the most popular one [21].

## 3. THE ARCHITECTURE SELECTION METHOD BASED ON TABU SEARCH

When artificial neural networks are used for forecasting, determining the best architecture which gives the best forecasting accuracy is an important issue. Each of possible architectures and their performance criterion values can be considered as solutions and objective function values, respectively. Thus, the problem of finding the best architecture can be interpreted as an optimization problem [6]. Aladag [2] solved this optimization problem by using tabu search algorithm. He improved a tabu search algorithm to determine the best architecture which gives the most accurate forecasts. In his algorithm, he focused on the feed forward neural networks and claimed that his algorithm can also be extended to other type of neural networks such as feed-back neural networks.

Aladag's [2] architecture selection method was proposed determine the best feed forward neural network architecture that includes one hidden layer and one neuron in the output layer which is shown in Fig. **1**. The tabu search algorithm proposed by him starts with an initial solution. This solution is generated using a starting pool strategy [2]. Why the starting pool strategy is used in Aladag's [2] algorithm can be explained as follows [6]:

The tabu search algorithm is a meta heuristic method and the starting conditions such as starting solution and initial state of the memories are important for reaching better global solutions. In order to make good choices in the beginning, a strategy called "starting pool" proposed by Aladag [2] is utilized. In this strategy, a specified number of solutions are examined and the information gathered from this examination is used to generate starting solution and the initial state of the memories. Therefore, the number of examined solutions in the starting pool is a parameter for the proposed algorithm. Detailed information about the starting pool strategy can be found in [2]. After the starting conditions are determined by employing the starting pool strategy, the proposed tabu search algorithm can be used to find a good architecture which produces accurate forecasts.

As mentioned before, the tabu search algorithm is used to determine a feed forward neural network architecture including one hidden layer and one output neuron given in Fig. **1**. Therefore, this algorithm will find the number of inputs and the number of neurons in the hidden layer. The elements of tabu search algorithm proposed by Aladag [2] are briefly given in below.

Let vector $x$ represents a solution whose elements are the numbers of neurons in the input and the hidden layers. The number of output neuron is always equal to one so there is no need to include this number in the solution vector. Thus, each solution $x$ has two elements. The solution space $X$ is a set of solutions. The solution space includes all of the possible ANN architectures. The objective function is root mean square error (RMSE) which measures the difference between the actual and the forecasted values obtained from a solution $x$. RMSE value is computed over the test set. The objective function can be defined by

$$f(x) = \left( \frac{\sum_{i=1}^{T}(y_i - \hat{y}_i)^2}{T} \right)^{1/2} \tag{1}$$

where $y_i$ is the actual value, $\hat{y}_i$ is the predicted value forecasted by using the solution $x$, and $T$ represents the size of test data. Since the objective function is RMSE, the tabu search algorithm will try to minimize RMSE value. In other words, the algorithm will try to find an architecture that produces the most accurate forecasts. The detailed information about the all elements of this tabu search algorithm can be found in Aladag [2].

In tabu search algorithm proposed by Aladag [2], diversification and intensification strategies are also used to improve the algorithm's performance. These strategies can be given as follows [6]:

*Diversification strategy*: In order to generate a diversification effect, restarting strategy is employed. When a new starting solution is being generated, the information obtained from the frequency based memories is used. The inputs and neurons in the hidden layer with large number in the memories are penalized to increase the selection probability of inputs and neurons with small number. After a pre determined number of iterations, which is called diversification parameter for the proposed algorithm, if a solution better than the best global solution obtained so far is not found, the algorithm will start to explore another region by restarting from a new starting solution. Also, when all neighbors of a current solution cannot be accepted, the next current solution is generated by using frequency based memory in a similar way. Therefore, when the algorithm is stuck in a region, the search is directed to unexplored regions by utilizing the diversification strategy.

*Intensification strategy*: An intensification strategy is designed by utilizing critical event memory. The solutions are vectors consist of only two units which are represents the numbers of inputs and neurons in the hidden layer. Thus, these vectors do not need too much memory space to be saved. All examined current solutions are considered as critical events and are kept by using a critical event memory. After the stopping criterion is satisfied and the algorithm is terminated, three solutions in the critical event memory which have the best objective function values are examined again. The search focuses on these solutions and a kind of strategic oscillation strategy is applied to the weights of the networks represented by these solutions. By utilizing the strategic oscillation, weights values are changed. It is try to obtain better weight values for these networks in order to get better forecasts. The number of generated neighbors obtained by changing the weight values is the intensification parameter of the search algorithm.

After the definitions of the elements of the tabu search algorithm are briefly introduced, the pseudo code of the architecture selection algorithm can be given as follows [6]:

A)   Run the starting pool strategy.

B)   Set the best solution obtained from the starting pool strategy as the best global solution.

C)   Initialize the memories.

D)   Run the tabu search algorithm.

**Step 1.** Generate the starting solution by using the information obtained from the starting pool strategy and begin from this solution.

**Step 2.** Use the neighborhood structure to generate a candidate list for the current solution.

**Step 3.** Set the solution with the best objective function value in the candidate list as the current neighbour:

i)    If the current neighbor is not tabu, set this one as the next current solution then go to Step 4. If the current neighbor is tabu, go to Step 3 ii.

ii)   If the tabu solution satisfies the aspiration criterion, set this solution as the next current solution and go to Step 4. Otherwise, go to Step 3 iii.

iii)  Set the other neighbor with the best objective function value among the remains candidate solutions as the current neighbor and go to Step 3 i. If all of the candidate solutions are examined, go to Step 4.

*Step 4.* If all of the examined solutions in the previous step could not be set as the next current solution, then generate a solution by using the information obtained from frequency based memory and set this solution as the next current solution and go to Step 5. Otherwise, go to Step 5 directly.

*Step 5.* Update the memories according to selected current solution:

    *i)*     Update tabu list by adding the current solution.

    *ii)*    Update frequency based memories.

    *iii)*   Update the critical event memory by adding the current solution.

*Step 6.* If the objective function value of the current solution is better than the best value found so far, set this current solution as the best global solution, set diversification counter value as zero, and go to Step 7. Otherwise, increase the diversification counter by 1 and go to Step 7.

*Step 7.* If the diversification counter is smaller than the specified diversification parameter, then go to Step 8. Otherwise, run the diversification strategy to generate the next current solution, and go to Step 8.

*Step 8.* If the pre determined maximum iteration bound is reached, go to Step 9. Otherwise, go to Step 2.

*Step 9.* For three solutions in the critical event memory with the best objective values, run the intensification strategy. If a solution better than the best global solution found so far is obtained as a result of the intensification strategy, then set this solution as the best global solution.

The neighborhood structure which directly affects the performance of tabu search algorithm is a vital issue. Detailed information about the neighborhood structure of the algorithm given above can be found in [18].

## 4. THE IMPLEMENTATION

The architecture selection method based on tabu search is applied to the time series of air pollution of Ankara. Ankara is the capital city of Turkey and the second largest city in the country after Istanbul. The population of the city is around 3.5 million. The time series includes monthly sulfur dioxide ($SO_4$) quantities (microgram/$m^3$) in the air. The graph of the air pollution time series with 83 monthly observations between January 1996 and November 2002 is shown in Fig. **2**.

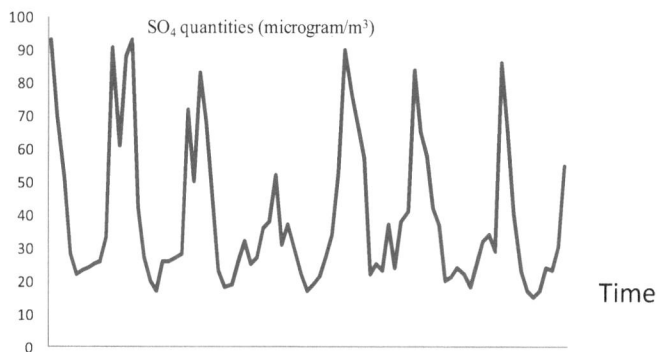

**Figure 2:** The air pollution time series.

While the best architecture that gives the most accurate forecasts is being searched, the other elements of the ANN are fixed like in [2]. The logistic activation function is used in all of the neurons of networks. The last 5 observations are taken as the test set and the rest of the observations are used for the training set. When networks are trained, Levenberg Marquardt algorithm [23] is employed as training algorithm.

Minimum and maximum numbers of inputs for the solution space are 1 and 15, respectively. Same values are used for the number of neurons in the hidden layer. The rest of parameters of the tabu search algorithm are taken as in Aladag [2]. Minimum and maximum numbers of inputs for the starting pool are 1 and 3, respectively. The same values are also used for the number of neurons in the hidden layer. In other words, both the number of inputs and the number of neurons in the hidden layer are changed from 1 to 3. Thus, 9 solutions are examined in the starting pool strategy. Tabu tenure and maximum iteration bound are taken as 20 and 50 iterations, respectively. The repeat parameter for each solution is 8. Finally, the diversification and the intensification parameters are chosen as 8 and 40, respectively. (See Aladag [2] for detailed information about the parameters of the used tabu search algorithm).

When the air pollution time series are forecasted by feed forward neural networks, two methods which are trial and error and the approach proposed by Aladag [2] are utilized in the architecture selection process. For each architecture selection method, the time series is analyzed with feed forward neural networks for 100 times. The obtained average RMSE values calculated over the test sets are summarized in Table **1**.

**Table 1:** The obtained results

| Architecture selection method | Average RMSE |
|---|---|
| Trial & Error | 3.34 |
| Aladag's [2] algorithm | 2.01 |

According to Table **1**, it is clearly seen that using architecture selection method proposed by Aladag [2] produces more accurate forecasts than those obtained from trial and error method when the air pollution time series is forecasted by feed forward neural networks. The best architectures determined by using these architecture selection methods and corresponding RMSE values are presented in Table **2**. The real values in the test set and the forecast values obtained from the architectures determined by using both architecture selections are summarized in Table **3**.

**Table 2:** The best architectures

| Architecture selection method | The best architecture | RMSE value |
|---|---|---|
| Trial & Error | 2-6-1 | 1.89 |
| Aladag's [2] algorithm | 7-5-1 | 1.37 |

**Table 3:** The real values and the forecasts

| Real values | Aladag's [2] method | Trial & Error method |
|---|---|---|
| 17 | 15.00 | 19.25 |
| 24 | 21.82 | 25.93 |
| 23 | 22.56 | 23.88 |
| 30 | 30.32 | 27.30 |
| 55 | 55.59 | 53.97 |

It is seen from Table **2** that a better architecture can be obtained which gives more accurate forecasts when Aladag's [2] algorithm is employed for the architecture selection. The architecture 7-5-1 in which 7, 5 and 1 neurons are used in input, hidden and output layers, respectively is found as the best architecture when Aladag's [2] algorithm is used. The obtained forecasting results are also examined visually. The graph of the real and the forecast values obtained from the architecture 2-6-1, which is found by using trial and error method, is given in Figs. **3** and **4** gives the graph of the real and the forecast values obtained from the architecture 7-5-1, which is found by using Aladag's [2] algorithm. When the graphs are examined, it is obvious that better forecasts are obtained from the architecture 7-5-1 which is determined by using the architecture selection method based on tabu search algorithm.

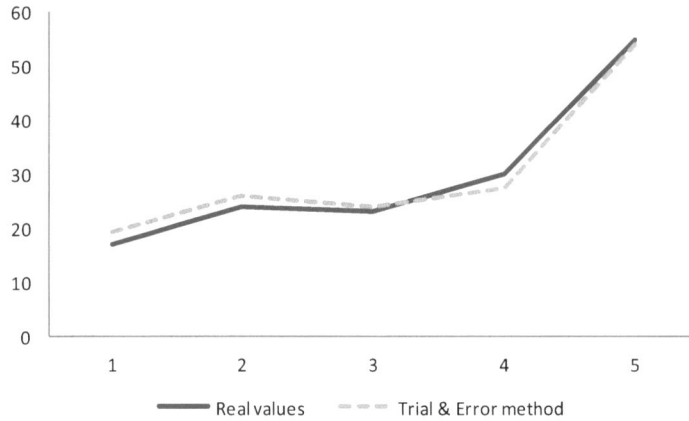

**Figure 3:** The graph of the real and the forecast values obtained from trial and error method.

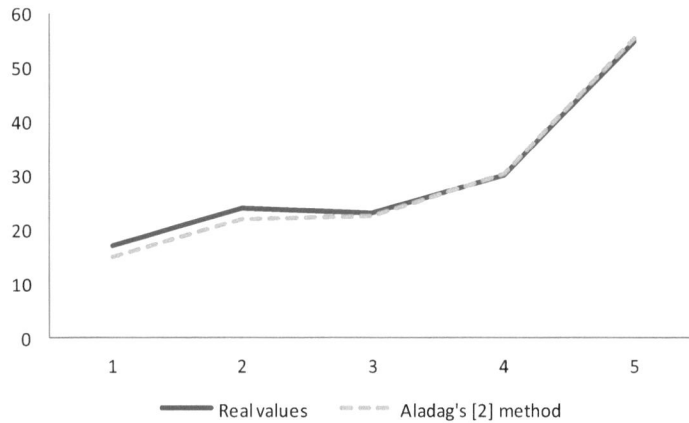

**Figure 4:** The graph of the real and the forecast values obtained from Aladag's [2] algorithm.

## 5. CONCLUSION

It is a well known fact that forecasts obtained from artificial neural networks can be more accurate than those obtained from conventional time series forecasting method. In various fields, therefore, there are many forecasting studies in which artificial neural networks are employed. On the other hand, there are some problems with using this method in forecasting problem. There are no general rules to determine the elements of this approach which have important effect on the forecasting performance.

One of the important issues is to determine the best architecture which gives the most accurate forecasts when artificial neural networks are used to forecast time series. Although there have been some systematic approaches to find the best architecture, the most preferred method has been trial and error method in the applications. One of the systematic approaches proposed to determine best architecture is the architecture selection approach proposed by Aladag [2]. He considered the architecture selection problem as an optimization problem in which the forecasting error is tried to be minimized and improved a tabu search algorithm to solve this problem.

In this study, the air pollution in Ankara time series is forecasted by using feed forward neural networks and the approach proposed by Aladag [2] is employed to find the best architecture. Also the trial and error method is used to determine the best architecture for comparison. As a result of the comparison, it is clearly seen that the most accurate forecasts are obtained when Aladag's [2] algorithm is used to find the best architecture.

## REFERENCES

[1]     P. Leahy, G. Kiely, and G. Corcoran, "Structural optimisation and input selection of an artificial neural network for river level prediction," *J Hydrol*, vol. 355, pp. 192-201, 2008.

[2]     C.H. Aladag, "Using tabu search algorithm in the selection of architecture for artificial neural networks", Ph.D. thesis, Hacettepe University, Turkey, 2009.

[3]     C.H. Aladag, E. Eğrioğlu, and C. Kadilar, "Forecasting nonlinear time series with a hybrid methodology, *Appl Math Lett*, vol. 22, pp. 1467-1470, 2009.

[4]     C.H. Aladag, E. Eğrioğlu, S. Gunay, and M.A. Basaran, "Improving weighted information criterion by using optimization," *J Comput Appl Math*, vol. 233, pp. 2683-2687, 2010.

[5]     E. Eğrioğlu, C.H. Aladag, and S. Gunay, "A new model selection strategy in artificial neural network" *Appl Math Comput*, vol. 195, pp. 591-597, 2008.

[6]     C.H. Aladag, "A new architecture selection method based on tabu search for artificial neural networks, *Expert Syst Appl*, vol. 38, pp. 3287-3293, 2011.

[7]     J. Siestema, and R. Dow, "Neural net pruning: why and how?", In: Proceedings of the IEEE International Conference on Neural Networks, 1988, pp. 325-33.

[8]     Roy, L.S. Kim, and S. Mukhopadhyay, "A Polynomial time algorithm for the construction and training of a class of multilayer perceptrons," *Neural Networks*, vol. 6, pp. 535-545, 1993.

[9]     N. Murata, S. Yoshizawa and S. Amari, "Network information criterion determining the number of hidden units for an artificial neural network model," *IEEE Trans Neural Netw*, vol. 5, pp. 865-872, 1994.

[10]    T.F. Rathbun, S.K. Rogers, M.P. DeSimo, and M.E. Oxley, "MLP iterative construction algorithm," *Neurocomputing*, vol. 17, pp. 195-216, 1997.

[11]    S. Buhamra, N. Smaoui, and M. Gabr, "The box-jenkins analysis and neural networks: prediction and time series modelling," *Appl Math Model*, vol. 27, pp. 805-815, 2003.

[12]    H.C. Yuan, F.L. Xiong, and X.Y. Huai, "A method for estimating the number of hidden neurons in feed-forward neural networks based on information entropy," *Comput Electron Agr*, vol. 40, pp. 57-64, 2003.

[13]    M. Dam, and D.N. Saraf, "Design of neural networks using genetic algorithm for on-line property estimation of crude fractionators products," *Comput Chem Eng*, vol. 30, pp. 722–729, 2006.

[14]    J. Zeng, H. Guo, and Y. Hu, "Artificial neural network model for identifying taxi gross emitter from remote sensing data of vehicle emission," *J Environ Sci*, vol. 19, pp. 427–431, 2007.

[15]    Durbin, S. Dudoit, and M.J Van der Laan, "A deletion/substitution/addition algorithm for classification neural networks, with applications to biomedical data," *J Stat Plan Infer*, vol. 138, pp. 464-488, 2008.

[16]    C.H. Aladag, E. Eğrioğlu, and S. Gunay, "A new architecture selection strategy in solving seasonal autoregressive time series by artificial neural networks," *Hacet J Math Stat*, vol. 37, no. 2, pp. 185-200, 2008.

[17]    P.P. Balestrassi, E. Popova, A.P. Paiva, and J.W.M. Lima, "Design of experiments on nn training for nonlinear time series forecasting," *Neurocomputing*, vol. 72, no. 4-6, pp. 1160-1178, 2009.

[18]    C.H. Aladag, G. Hocaoglu, and M.A. Basaran, "The effect of neighborhood structures on tabu search algorithm in solving course timetabling problem," *Expert Syst Appl*, vol. 36, pp. 12349–12356, 2009.

[19]    M. Qi, and G. Zhang, "An investigation of model selection criteria for neural network time series forecasting," *Eur J Oper Res*, vol. 132, pp. 666-680, 2001.

[20]    S. Buhamra, N. Smaoui, and M. Gabr "The Box-Jenkins Analysis and Neural Networks: Prediction and Time Series Modelling," *Appl Math Model*, vol. 27, pp. 805-815, 2003.

[21]    G. Zhang, B.E. Patuwo, and Y.M. Hu, "Forecasting with artificial neural networks: the state of the art," *Int J Forecasting*, vol. 14, pp. 35-62, 1998.

[22]    S. Gunay, E. Eğrioğlu, and C.H. Aladag, *Introduction to single variable time series analysis*. Hacettepe University Press: Ankara, 2007.

[23]    K. Levenberg, "A method for the solution of certain non-linear problems in least squares," *Q Appl Math*, vol. 2, pp. 164–168, 1944.

# CHAPTER 12

## A Hybrid Forecasting Approach Combines SARIMA and Fuzzy Time Series

### Erol Eğrioğlu[1], Cagdas Hakan Aladag[2,*] and Ufuk Yolcu[1]

*[1]Ondokuz Mayis University, Faculty of Arts and Science, Department of Statistics, 55139, Samsun, Turkey and [2]Hacettepe University, Faculty of Science, Department of Statistics, 06800, Ankara, Turkey*

**Abstract:** Fuzzy time series, subjected to many scientific studies, have been used in forecasting in recent years. Due to their uncertainty, time series encountered in daily life should be perceived as fuzzy time series and analyzed by fuzzy time series methods. Instead of representing time series, which may have different values during the time they measured, by instantaneous value of each observation, representing a fuzzy set which may contain several values provides more information and thus more realistic analyses. In such a situation, forecasting problem of time series whose observations are fuzzy sets emerges. In the literature, there are several methods and algorithms proposed for forecasting these types of fuzzy time series. However, one can say that most of the observed fuzzy time series contain seasonal structures. From this stand point, using seasonal fuzzy time series forecasting methods in analyzing fuzzy time series containing seasonal relations would be effective in terms of both forecasting performance and explanation of the relation of the data contained in. This study aims to introduce a partial high order bivariate fuzzy time series forecasting method hybridized with Box-Jenkins method seasonal autoregressive integrated moving average model (SARIMA), one of the conventional time series analysis methods used in forecasting seasonal time series, and its advantages. For this purpose, two real data are analyzed using this seasonal fuzzy time series forecasting method and results are evaluated with certain fuzzy and conventional seasonal time series methods.

**Keywords:** Bivariate fuzzy time series, Forecasting, High order, Seasonal fuzzy time series.

## 1. INTRODUCTION

Fuzzy time series methods are effective methods in forecasting time series which contain uncertainty in their observations and which can be frequently encountered in real life. Not having constraints which are seen in other methods, such as number of observation and model assumption, have attracted attention on fuzzy time series methods.

Fuzzy set theory was first introduced by Zadeh [1]. Then, Song and Chissom proposed the fuzzy time series [2, 3] based on fuzzy set theory. Fuzzy time series analysis method consists of three steps as fuzzifying the observations, identifying fuzzified relations and defuzzifying. While Song and Chissom used matrix operations and compound processes [2-4], Chen proposed an easy method based on fuzzy logical relation tables [5]. Moreover, Huarng and Yu [6] proposed a model in which fuzzy relations are determined by artificial neural networks. In the literature, all the methods proposed by Sullivan and Woodal [7], Hwang, Chen and Lee [8], Chen and Hwang [9], Huarng and Yu [10] and Yu [11, 12] include first order fuzzy time series forecasting models. However, analyzing fuzzy time series with first order models, gives insufficient results as they contain high order relations. Therefore, Chen [13] proposed a new model to analyze fuzzy time series that contains high order relations. All lagged variables belong to fuzzy time series are present in the high order method proposed by Chen [13]. For instance, when a fourth order fuzzy time series forecasting model is created, the model includes all the lagged variables in the first, second, third and fourth orders. For seasonal time series, model order should be equal to the period. However, this adds pure lagged variables to the model and unnecessarily increases the number of input in the model. Additionally, Song [14] proposed a new model where lagged variable $F(t-m)$ is the input and $F(t)$ is the output to analyze a seasonal time series whose period is $m$. It is a fact that this method will fail to analyze seasonal time series containing more complex relations.

*Address correspondence to Cagdas Hakan Aladag: Hacettepe University, Faculty of Science, Department of Statistics, 06800, Ankara, Turkey; E-mail: chaladag@gmail.com

In order to eliminate this problem, Eğrioğlu *et al.* [15] proposed a model including seasonal autoregressive and seasonal moving average (MA) terms in analyzing fuzzy seasonal time series. Since both time series ($X_t$) and residuals series ($a_t$) are included in the model, the method proposed by Eğrioğlu *et al.* [15] has bivariate structure. Bivariate fuzzy time series method was first proposed by Huarng [16], Hsu, Tse, and Wu [17] and Yu and Huarng [18]. In his method, Huarng [16] used only first order lagged variables. On the other hand, Yu and Huarng used feed forward artificial neural networks in determining fuzzy relations.

In the method proposed by Eğrioğlu *et al.* [15], time series is primarily determined by Box-Jenkins method with SARIMA model. Then, partial high order, bivariate fuzzy time series using the inputs of determined SARIMA model is created. Determination of fuzzy relations is done by using artificial neural networks. The advantages of this model are as follows:

- It establishes a high order model in forecasting seasonal fuzzy time series.

- Model order is determined systematically.

- It is the first method in the literature including the term of MA.

- It improves the forecasting performance.

The method proposed by Eğrioğlu *et al.* [15] and emphasized in this study is applied to the time series of "the average of sulfur dioxide ($SO_2$) in Samsun province between January 2004 and December 2008 (SAMSO)" and "the number of foreign tourist arriving in Turkey between January 1998 and December 2008 (TOURISM DATA)" and obtained results are compared with those in the other methods. In the second section, SARIMA models and artificial neural networks are introduced and basic definitions of fuzzy time series are given. The method and its algorithm proposed by Eğrioğlu *et al.* [15] are introduced in the third section. Fourth section deals with the application. In the last section, results obtained from the study are discussed considering the results of other methods in the literature.

## 2. RELATED METHODS

In this section, SARIMA models and artificial neural networks, the basis of method intended to be introduced, and basic definitions of fuzz time series are given.

### 2.1. SARIMA Models

Let $Z_t$ be a time series with $\mu$ mean. Then, the model is

$$\varphi(B)\Phi(B^s)(1-B)^d(1-B^s)^D(Z_t - \mu) = \theta(B)\Theta(B^s)a_t \qquad (1)$$

where,

$$\varphi(B) = (1 - \varphi_1 B - \cdots - \varphi_p B^p) \qquad (2)$$

$$\theta(B) = (1 + \theta_1 B + \cdots + \theta_q B^q) \qquad (3)$$

$$\Phi(B) = (1 - \Phi_1 B^s - \cdots - \Phi_P B^{sP}) \qquad (4)$$

$$\Theta(B) = (1 + \Theta_1 B^s + \cdots + \Theta_Q B^{sQ}) \qquad (5)$$

Box-Jenkins first proposed a method for SARIMA, expressed by $SARIMA(p,d,q)(P,D,Q)_s$ [19]. This method is known as Box-Jenkins method and consists of determination, estimation, diagnostic control and forecasting stages. Detailed information for SARIMA models and Box-Jenkins method can be obtained from [19].

## 2.2. Artificial Neural Networks

Artificial neural networks are synthetic networks that imitate biological neural networks. There are great differences between artificial neural networks and biological neural networks in terms of their architectures and abilities [20]. Artificial neural networks constitute mathematical models and known as convergence of function [21].

More recently, artificial neural networks, as an alternative to other methods, began to be used in some stages such as determination of relation in the analysis of fuzzy time series. In the studies [6] and [18], feed forward artificial neural networks were used for the determination of fuzzy relations in fuzzy time series. The determination of fuzzy relations with artificial neural networks removes the complexity in defining fuzzy logical relation and group relation tables and improves forecasting accuracy.

There are three components directing function of artificial neural networks. These are architecture structure, learning algorithm and activation function.

*Architecture structure:* Multi layer feed forward artificial neural network architecture used in artificial neural networks literature, consists of three layers. These are, input layer, hidden layer (or layers) and output layer. Layers consist of units called neuron (node). Accurate determination of architecture is achieved by deciding the number of neuron in layers. In artificial neural networks, neurons are interconnected *via* weights. In feed forward networks, these connections are unidirectional. And, there is no connection between the units in the same layer.

While multi layer artificial neural network architectures have single input and output layer, they may contain more than one hidden layers. The determination of architecture means determination of neuron number in the layers. A multi layer feed forward artificial neural network architecture is presented in Fig. **1**.

In Fig. **1**, $w_{ij}$ represents the weight between neuron $i$ in input layer and neuron $j$ in hidden layer and $w_{jk}$, represents the weight between neuron $j$ in hidden layer and neuron $k$ in hidden layer. As there is only one neuron in the figure, $k$ index shows this single neuron. In another architecture, a direct connection can be established between neuron in input layer and neuron(s) in output layer.

*Learning Algorithm:* There are several learning algorithms used in the determination of weights in artificial neural networks. Two of the most widely used algorithms are Back Propagation and Levenberg Marquardt learning algorithms. Detailed information on Back Propagation and Levenberg Marquardt learning algorithm can be obtained from [22-28] respectively.

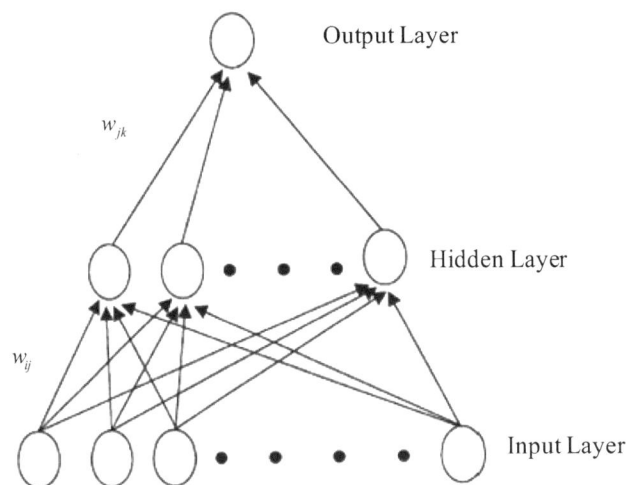

**Figure 1:** Multilayer feed forward artificial neural network with one output neuron.

*Activation Function:* Activation function enables curvilinear matchup between input and output units. The accurate selection of activation function affects performance of networks substantially. When the selected activation function is not linear, gradient parameter must be determined. Gradient parameter is a factor that plays an important role in reaching accurate conclusion. The expression given as scalar product of weight and input vector is called *net* activation value.

$$net = w^t x \tag{6}$$

In this case, domain of activation function will be the set of *net* activation values. Different activation functions are used for different types of artificial neural networks. However, different activation functions can be used in neurons, even if they are in the same network. Activation function can be unipolar, bipolar and linear. For unipolar *f(net)* activation function, $net \in R$ or $f(net) \in (0,1)$ whereas for bipolar *f(net)* activation function $net \in R$ or $f(net) \in (-1,1)$. Nevertheless, activation function may be discrete and constant. Obtained neuron output values with respect to selected activation function are given in Table **1**. In this table, *o* represents the output value of neuron.

One of the most preferred activation functions in the literature is logistic function given below.

$$f(net) = \frac{1}{1+\exp(-\lambda\ net)} \tag{7}$$

where $\lambda$ represents gradient parameter.

**Table 1:** Neuron output value according to activation function type

| Type of Activation Function | | Neuron output value |
| --- | --- | --- |
| Unipolar | Continuous | $o \in (0,1)$ |
| | Discrete | $o \in \{0,1\}$ |
| Bipolar | Continuous | $o \in (-1,1)$ |
| | Discrete | $o \in \{-1,1\}$ |

### 2.3. Fuzzy Time Series

Basic concepts related to fuzzy time series approaches proposed by Song and Chissom can be given as follows [2, 3];

**Definition 1.** Let $Y(t)$, $t = ...,0,1,2,...$ a subset of real numbers. After defining universe of discourse, the new time series $F(t)$ consisting of $A_i$ is called a fuzzy time series.

**Definition 2.** If fuzzy time series $F(t)$ contain seasonality, first order seasonal fuzzy time series forecasting model is

$$F(t\text{-}1) \rightarrow F(t) \tag{8}$$

here, *m* represents period.

**Definition 3.** If fuzzy time series $F(t)$ is affected by $F(t\text{-}1)$, $F(t\text{-}2)$, ..., $F(t\text{-}n)$ delayed time series, fuzzy logical relation can be expressed as;

$$F(t\text{-}n),.... F(t\text{-}2), F(t\text{-}1) \rightarrow F(t) \tag{9}$$

and called $n^{th}$ order fuzzy time series forecasting model.

## 3. THE HYBRID APPROACH COMBINES SARIMA AND FUZZY TIME SERIES

The following definition concerning the model architecture proposed by Eğrioğlu *et al.* can be given [15].

**Definition 4.** Let two fuzzy time series be $F(t)$ and $G(t)$. If $F(t)$ fuzzy time series is affected by $F(t-m_1),...,F(t-m_{k-1}),F(t-m_k),G(t-n_1),...,G(t-n_{l-1}),G(t-n_l)$ delayed fuzzy time series, then fuzzy logic relation is expressed as;

$$F(t-m_1),...,F(t-m_{k-1}),F(t-m_k),G(t-n_1),...,G(t-n_{l-1}),G(t-n_l) \rightarrow F(t) \tag{10}$$

is called $(k,l)^{\text{th}}$ order partial univariate fuzzy time series forecasting model in which outputs are determined by SARIMA model. Here, $m_i$ $(i=1,2,..,k)$ and $n_j$ $(j=1,2,..,l)$ are integer and as $1 \le m_1 < ... < m_k$, $1 \le n_1 < ... < n_l$.

The algorithm of hybrid approach proposed in [15] can be summarized as follows:

**Step 1.** Appropriate SARIMA model in accordance with Box-Jenkins procedure for time series is determined. Residuals $(a_t)$ obtained from SARIMA model are calculated [19].

**Step 2.** Universe of discourse and sub intervals for residuals obtained from time series and SARIMA are defined. Where $D_{\min}$ and $D_{\max}$ are the smallest and biggest observations of the original data respectively, universe of discourse can be defined as; $U = \left[D_{\min} - D_1, D_{\max} + D_2\right]$. Here, $D_1$ and $D_2$ are two positive numbers.

**Step 3.** Fuzzy sets are determined depending on universe of discourse and sub intervals. Based on $U$ and $V$ universe of discourse and partitioning, $A_1, A_2,...,A_{k_1}$ and $B_1, B_2,...,B_{k_2}$ linguistic varieties for time series and residuals can be defined as follows;

$$
\begin{aligned}
A_1 &= a_{11}/u_1 + a_{12}/u_2 + ... + a_{1n_1}/u_{n_1} & B_1 &= b_{11}/v_1 + b_{12}/v_2 + ... + b_{1n_2}/v_{n_2} \\
A_2 &= a2_{11}/u_1 + a_{22}/u_2 + ... + a_{2n_1}/u_{n_1} & B_2 &= b_{21}/v_1 + b2_{12}/v_2 + ... + b_{2n_2}/v_{n_2} \\
&\qquad\vdots & &\qquad\vdots \\
A_{k_1} &= a_{k_11}/u_1 + a_{k_12}/u_2 + ... + a_{k_1n_1}/u_{n_1} & B_{k_2} &= b_{k_21}/v_1 + b_{k_22}/v_2 + ... + b_{k_2n_2}/v_{n_2}
\end{aligned} \tag{11}
$$

Here, $a_{ij}$, $u_i$ are membership values of partition, and defined as; $a_{ij} \in [0,1]$, $1 \le i \le k_1$ and $1 \le j \le n_1$. Similarly, $b_{ij}$, $v_i$ are membership values of partition and defined as; $b_{ij} \in [0,1]$, $1 \le i \le k_2$ and $1 \le j \le n_2$.

**Step 4.** Fuzzify time series.

$F(t)$ represents fuzzy time series and $G(t)$ represents fuzzy residuals series obtained from SARIMA.

**Step 5.** Model order $(k,l)$ and $m_1,...,m_k$ and $n_1,...,n_l$ values are determined. For example, let model order be $k=5$ and $l=2$, then the model is,

$$F(t-1),F(t-2),F(t-12),F(t-13),F(t-14),G(t-1),G(t-12) \rightarrow F(t) \tag{12}$$

where $m_1=1$, $m_2=2$, $m_3=12$, $m_4=13$, $m_5=14$, $n_1=1$, $n_2=12$, $F(t)$, fuzzy $X_t$ ve $G(t)$, fuzzy $a_t$.

**Step 6.** Fuzzy relations are determined.

Fuzzy relations are established by using $F(t-m_1),...,F(t-m_{k-1}),F(t-m_k)$ belonging to fuzzy time series and $G(t-n_1),...,G(t-n_{l-1}),G(t-n_l)$ belonging to fuzzy errors as delayed variables and target value as $F(t)$. In this step, feed forward artificial neural network is trained according to input and target values.

***Step 7.*** Forecasts are obtained.

In the given model, in consequence of training artificial neural network, if the inputs of network are

$F(t+k-m_1),...,F(t+k-m_{k-1}),F(t+k-m_k)$ $G(t+k-n_1),...,G(t+k-n_{l-1}),G(t+k-n_l)$ and target is $F(t)$, then

$\hat{F}(t+k)$ as the output of the network, will be the fuzzy forecast. Artificial neural network architecture with these properties is given in Fig. **2**.

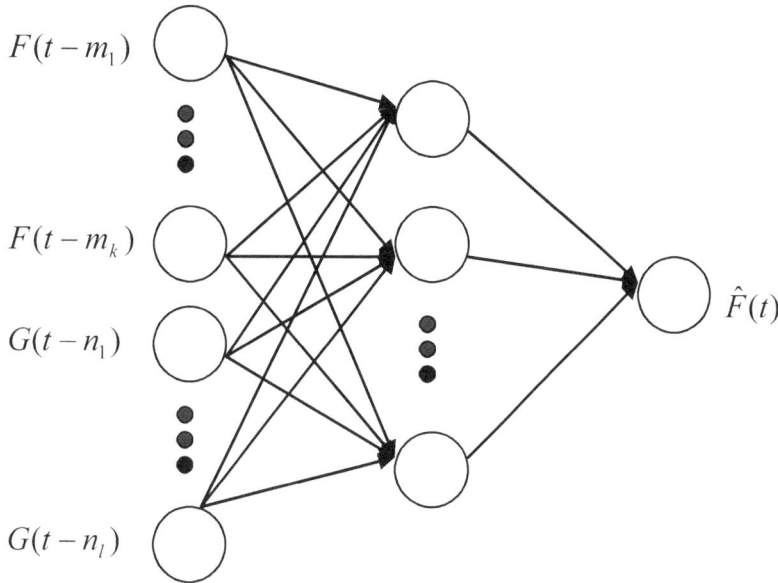

**Figure 2:** ANN architecture.

***Step 8.*** Defuzzifying process is applied using centralization method.

Centralization is used in defuzzification. If fuzzy forecast is obtained as $A_j$, then defuzzyfied forecast will be the midpoint which has got the biggest membership value in $A_j$ fuzzy set.

## 4. THE IMPLEMENTATION

The method proposed by Eğrioğlu *et al.* [15] and emphasized in this study is applied to the time series of the averages of sulfur dioxide (SO2) in Samsun province between January 2004 and December 2008 (SAMSO), which is presented in Fig. **3** and the number of foreign tourist arriving in Turkey between January 1998 and December 2008 (TOURISM DATA). Root mean squared error (RMSE) and mean absolute percentage error (MAPE) values are used in the evaluation of obtained results. The related formulas are given below.

$$RMSE= \sqrt{\frac{\sum_{i=1}^{n}(forecast_t - real_t)^2}{n}} \tag{13}$$

$$MAPE = \frac{1}{n}\sum_{i=1}^{n}\left|\frac{forecast_t - real_t}{real_t}\right| \tag{14}$$

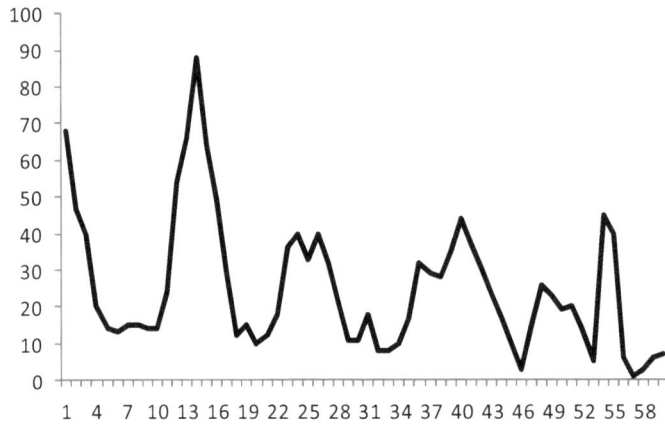

**Figure 3:** The average of sulfur dioxide ($SO_2$) in Samsun province between January 2004 and December 2008.

## 4.1. SAMSO Applicaton

The model algorithm consisting of 8 steps for the time series given in Fig. **3** is as follows:

***Step 1.*** Appropriate SARIMA model to related time series is determined using Box-Jenkins method and inputs of model are defined by calculating residuals $(a_t)$ obtained from SARIMA model. $SARIMA(0,1,2)(1,1,0)_{12}$ is found the most appropriate model according to Box-Jenkins method. If $f$ is the linear function, this model can be written as follows:

$$SARIMA(0,1,2)(2,1,0)_{12} : X_t = f(X_{t-1}, X_{t-12}, X_{t-13}, X_{t-24}, X_{t-25}, a_{t-1}, a_{t-2}) \tag{15}$$

where $m_1 = 1$, $m_2 = 12$, $m_3 = 13$, $m_4 = 24$, $m_5 = 25$, $n_1 = 1$, $n_2 = 2$ and $F(t)$ is fuzzy $X_t$ and $G(t)$ is fuzzy $a_t$.

***Step 2.*** In accordance with the output of SARIMA model, model order $(5,2)$ and partial high order bivariate forecasting model $k = 5$ *and* $l = 2$ is expressed as:

$$F(t-1), F(t-12), F(t-13), F(t-24), F(t-25), G(t-1), G(t-2) \rightarrow F(t) \tag{16}$$

***Step 3.*** Universe of discourse and sub intervals are defined for residuals obtained from time series and SARIMA.

The smallest and biggest values of times series are $D_{min} = 1$ and $D_{max} = 88$ and the smallest and biggest values of residuals are $D_{(art)min} = -37.84$ and $D_{(art)max} = 43.02$, respectively. In this case, universe of discourse for time series and residuals can be defined as $U = [0,100]$ and $V = [-40, 60]$, respectively.

In obtaining sub intervals of defined universe of discourse, the interval length for each series is defined as 2.5, 5, 10, 15 and 20. When the length of interval is defined as 20, sub intervals for time series and residuals are obtained as follows:

$$u_1 = [0, 20] \qquad u_2 = [20, 40] \quad v_1 = [-40, -20] \quad v_2 = [-20, 0]$$
$$u_3 = [40, 60] \qquad u_4 = [60, 80] \quad v_3 = [0, 20] \qquad\qquad v_4 = [20, 40]$$
$$u_5 = [80, 100] \qquad\qquad\qquad\qquad v_5 = [40, 60]$$

***Step 4.*** Fuzzy sets are determined depending on universe of discourse and sub intervals.

For example, when the length of interval for both time series and residuals defined as 20, $A_1, A_2, ..., A_5$ and $B_1, B_2, ..., B_5$ linguistic varieties for time series and residuals can be defined as follows;

$$A_1 = 1/u_1 + 0.5/u_2 + 0/u_3 + 0/u_4 + 0/u_5$$
$$A_2 = 0.5/u_1 + 1/u_2 + 0.5/u_3 + 0/u_4 + 0/u_5$$
$$A_3 = 0/u_1 + 0.5/u_2 + 1/u_3 + 0.5/u_4 + 0/u_5 \qquad (17)$$
$$A_4 = 0/u_1 + 0/u_2 + 0.5/u_3 + 1/u_4 + 0.5/u_5$$
$$A_5 = 0/u_1 + 0/u_2 + 0/u_3 + 0.5/u_4 + 1/u_5$$

$$B_1 = 1/v_1 + 0.5/v_2 + 0/v_3 + 0/v_4 + 0/v_5$$
$$B_2 = 0.5/v_1 + 1/v_2 + 0.5/v_3 + 0/v_4 + 0/v_5$$
$$B_3 = 0/v_1 + 0.5/v_2 + 1/v_3 + 0.5/v_4 + 0/v_5 \qquad (18)$$
$$B_4 = 0/v_1 + 0/v_2 + 0.5/v_3 + 1/v_4 + 0.5/v_5$$
$$B_5 = 0/v_1 + 0/v_2 + 0/v_3 + 0.5/v_4 + 1/v_5$$

***Step 5.*** Time series are fuzzified.

In this step, residuals obtained from both time series and SARIMA are matched with the fuzzy set which has the biggest membership value and are fuzzified.

***Step 6.*** Fuzzy relations are defined.

While delayed variables of fuzzy time series and fuzzy residuals $F(t-1), F(t-12), F(t-13), F(t-24), F(t-25), G(t-1), G(t-2)$ are input of artificial neural network, target value is $F(t)$. At this stage, the number of hidden layer units in feed forward artificial neural network varied between 1-12, thus analyses for 300 (5x5x12) different situations are done. Additionally, Levenberg-Marquardt learning algorithm is chosen as a learning algorithm and logistic activation function is used in input and output layer of artificial neural network.

***Step 7.*** Forecasts are obtained.

During the analysis, the last 6 observations of the data are defined as the test set and forecasted. As a result of training artificial neural network consisting of 7 inputs and one output mentioned in the previous step, forecasts for 300 different situations are obtained.

***Step 8.*** Defuzzification.

When all cases are analyzed, while the best case in terms of RMSE is obtained in the case in which interval length of time series is 5, interval length of residuals is 10 and hidden layer unit number is 8 with 2.3629, when considering MAPE values the best case was obtained in the case in which interval length of time series is 2.5, interval length of residuals is 10 and hidden layer unit number is 11 with 0.4199.

Additionally, related time series are analyzed with first order models proposed by Chen [5], seasonal first order model proposed by Song [14] and high order model proposed by Chen [13]. In these methods, interval lengths for times series are taken as 2.5, 5, 10, 15 and 20 as in the model proposed by Eğrioğlu *et al.* [15] and analysis are done using model orders varying between 2 and 12 in Chen's high order approach [13].

Table **2** and **3** show the best results obtained from [15] method and the results of other methods. When these tables are analyzed, it is seen that the partial high order bivariate hybrid approach proposed by Eğrioğlu *et al.* [15] gives more accurate forecasts in terms of RMSE and MAPE values in comparison with those in other methods. The graph of the real values of time series and forecasts of the best situations are given in Fig. **4**.

**Table 2:** The best RMSE values of the methods

| Method | Length of interval in time series | Length of interval in residuals | Hidden layer neurons number | RMSE | MAPE |
|---|---|---|---|---|---|
| Chen's method [5] | 15 | - | - | 17.9513 | 6.0253 |
| Song's method [14] | 2.5 | - | - | 10.4188 | 1.7453 |
| Chen's method [13] (seventh order) | 2.5 | - | - | 16.1581 | 5.5437 |
| Eğrioğlu et al.'s method [7] | 5 | 10 | 8 | **2.3629** | 0.4945 |

**Table 3:** The best MAPE values of the methods

| Method | Length of interval in time series | Length of interval in residuals | Hidden layer neurons number | RMSE | MAPE |
|---|---|---|---|---|---|
| Chen's method [5] | 15 | - | - | 17.9513 | 6.0253 |
| Song's method [14] | 2.5 | - | - | 10.4188 | 1.7453 |
| Chen's method [13] (second order) | 2.5 | - | - | 16.8879 | 4.0024 |
| Eğrioğlu et al.'s method [7] | 2.5 | 10 | 11 | 3.1721 | **0.4199** |

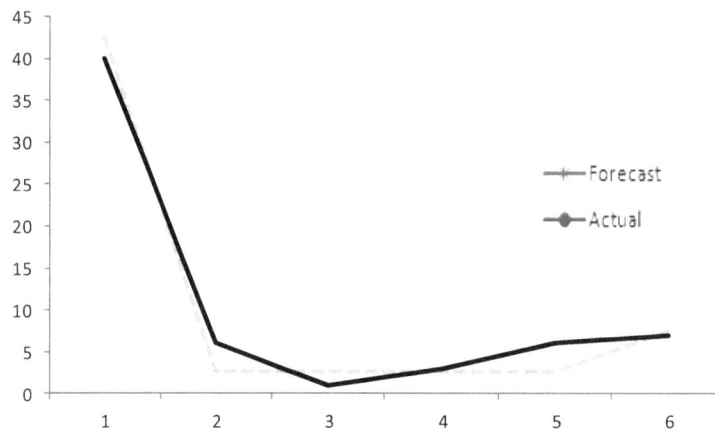

**Figure 4:** Forecast and real series graph of July 2008-December 2008 (test set).

### 4.2. Tourism Data Application

Time series of the number of foreign tourist arriving in Turkey between January 1998 and December 2008 which is called as TOURISM DATA in this study is used in the analysis. The graph of TOURISM DATA is shown in Fig. **5**. The last 12 observations of time series are used as test set. As detailed method algorithm is given in previous application, this time we will only emphasize model parameters.

In the first step of the TOURISM DATA analysis, the most appropriate model according to Box-Jenkins method is found as $SARIMA(1,0,2)(0,1,1)_{12}$. Let $f$ be a linear function, then this model can be written as follows;

$$SARIMA(1,0,2)(0,1,1)_{12} : X_t = f(X_{t-1}, X_{t-12}, X_{t-13}, a_{t-1}, a_{t-2}a_{t-14}, a_{t-13}, a_{t-14}) \qquad (19)$$

where $m_1 = 1$, $m_2 = 12$, $m_3 = 13$, $n_1 = 1$, $n_2 = 2$ $n_3 = 12$, $n_4 = 13$, $n_5 = 14$, $F(t)$ fuzzy $X_t$ and $G(t)$ fuzzy $a_t$.

In this case, the model order is (3,5) according to inputs of SARIMA model and partial high order bivariate forecasting model which is $k = 3$ and $l = 5$ is expressed as:

$$F(t-1), F(t-12), F(t-13), G(t-1), G(t-2)G(t-12), G(t-13), G(t-14) \rightarrow F(t) \qquad \textbf{(20)}$$

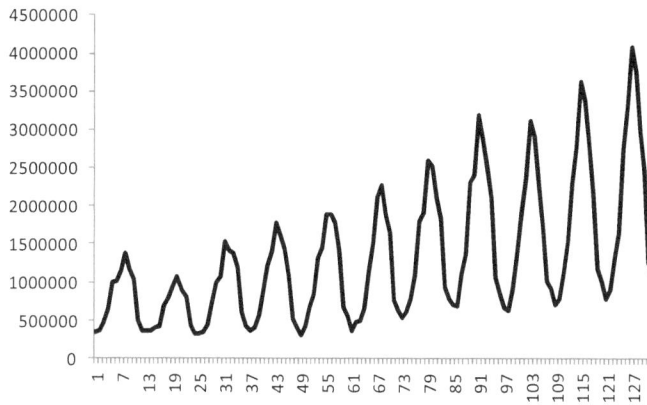

**Figure 5:** The number of tourist arriving in Turkey between January 1998 and December 2008.

In the analysis, the lengths of intervals for time series were taken as 200000, 250000, 300000, 350000, 400000, 450000, 500000, 550000 respectively and the lengths of intervals for residual time series were taken as 20000, 30000, 40000, 50000, 60000, 70000, 80000, 90000, 10000 respectively. Nevertheless, the unit number in the hidden layer of artificial neural network used in the determination of fuzzy relations varies between 1 and 16. While Levenberg-Marquardt learning algorithm is used as a learning algorithm during the determination of relations *via* artificial neural networks, logistic activation function is used in the hidden and the output layer of artificial neural network. In the light of this information, analyses are performed for 1296 different situations including all alternatives and based on 8 inputs and one output. In terms of performance evaluation, data is analyzed *via* first order model proposed by Chen [5], seasonal first order model proposed by Song [14] and high order model proposed by Chen [13].

At the end of the analyses, when all cases are analyzed while the best case in terms of RMSE was obtained in the case in which interval length of time series is 150000, interval length of residuals is 90000 and the hidden layer unit number is 3 with 228129, when considering MAPE values the best case iss obtained in the case in which interval length of time series is 30000, interval length of residuals is 90000 and the hidden layer unit number is 5 with 0.0809.

Tables **4** and **5** show the best results obtained from the method proposed by Eğrioğlu *et al.* [15] and the results of other methods. When these tables are examined, it is seen that the partial high order bivariate hybrid approach proposed by Eğrioğlu *et al.* [15] gives more accurate forecasts in terms of RMSE and MAPE values in comparison with those in other methods. The graph of the real values of time series and forecasts of the best situations are given in Fig. **6**.

**Table 4:** The best RMSE values of the methods

| Method | Length of interval in time series | Length of interval in residuals | Hidden layer neurons number | RMSE | MAPE |
|---|---|---|---|---|---|
| Chen's method [5] | 550000 | - | - | 646057 | 0.2544 |
| Song's method [14] | 550000 | - | - | 532778 | 0.2297 |
| Chen's method [13] (second order) | 500000 | - | - | 569496 | 0.2954 |
| Eğrioğlu *et al.*'s method [7] | 150000 | 90000 | 3 | 228129 | 0.0950 |

**Table 5:** The best MAPE values of the methods

| Method | Length of interval in time series | Length of interval in residuals | Hidden layer neurons number | RMSE | MAPE |
|---|---|---|---|---|---|
| Chen's method [5] | 250000 | - | - | 768324 | 0.2544 |
| Song's method [14] | 550000 | - | - | 532778 | 0.2297 |
| Chen's method [13] (second order) | 500000 | - | - | 569496 | 0.2954 |
| Eğrioğlu *et al.*'s method [7] | 300000 | 90000 | 5 | 288079 | 0.0809 |

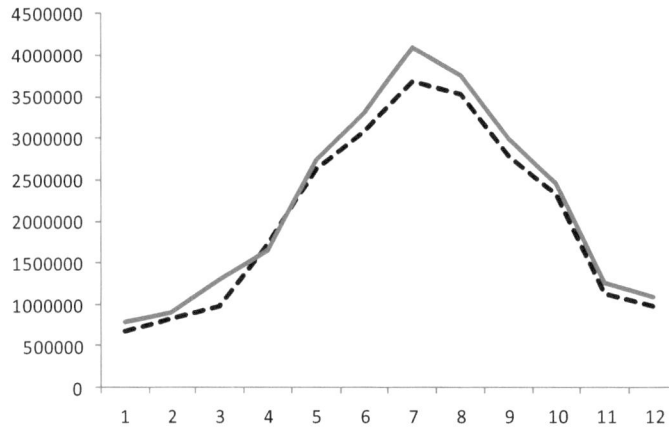

**Figure 6:** Graph of forecast-real series between January 2008-December 2008 (test set).

## 5. CONCLUSION

Due to their uncertainty, time series encountered in daily life should be perceived as fuzzy time series. Additionally, when compared with conventional methods, fuzzy time series is an effective method in forecasting problems due to its salient capabilities of tracking uncertainty in observation data. Most of the time series contain seasonal relations. In the analysis of this type of time series, methods which consider the seasonal relations should be applied. In such a case, selecting the model type becomes a problem. A partial high order bivariate hybrid method proposed by Eğrioğlu *et al.* [15] offers solutions to this problem and improves forecasting performance. The method emphasized in this study has some advantages. On the other hand, it determines the length of intervals subjectively. Methods which will eliminate basic problems at this stage and bring more systematic approaches are needed.

## REFERENCES

[1]   L.A. Zadeh, "Fuzzy Sets," *Inform and Control*, vol. 8, pp. 338-353, 1965.

[2]   Q.Song, and B.S. Chissom, "Fuzzy time series and its models," *Fuzzy Set Syst*, vol. 54, pp. 269-277, 1993.

[3]   Q. Song, and B.S. Chissom, "Forecasting enrollments with fuzzy time series - Part I," *Fuzzy Set Syst*, vol. 54, pp. 1-10. 1993.

[4]   Q. Song, and B.S. Chissom, "Forecasting enrollments with fuzzy time series - Part II," *Fuzzy Set Syst*, vol. 62, no. l, pp. 1-8, 1994.

[5]   S.M. Chen, "Forecasting enrollments based on fuzzy time-series," *Fuzzy Set Syst*, vol. 81, pp. 311-319, 1996.

[6]   K. Huarng, and H.K. Yu, "The application of neural networks to forecast fuzzy time series," *Physica A*, vol. 363, pp. 481-491, 2006.

[7]   J. Sullivan, and W.H. Woodall, "A comparison of fuzzy forecasting and Markov modeling," *Fuzzy Set Syst*, vol. 64, no. 3, pp. 279-293, 1994.

[8]   J.R. Hwang, S.M. Chen, and C.H. Lee, "Handling forecasting problems using fuzzy time series," *Fuzzy Set Syst*, vol. 100, pp. 217-228, 1998.

[9]    S.M. Chen, and J.R. Hwang, "Temperature prediction using fuzzy time series," *IEEE Trans Syst Man, Cybern B, Cybern*, vol. 30, no. 2, pp. 263-275, 2000.

[10]   K. Huarng, and H.K. Yu, "A type 2 fuzzy time series model for stock index forecasting," *Physica A*, vol. 353, pp. 445-462, 2000.

[11]   H.K. Yu, "A refined fuzzy time series model for forecasting," *Physica A*, vol. 346, pp. 657-681, 2005.

[12]   H.K. Yu, "Weighted fuzzy time series models for TAIEX forecasting," *Physica A*, vol. 349, pp. 609-624, 2005.

[13]   S.M. Chen, "Forecasting enrollments based on high order fuzzy time series," *Cybernet Syst*, vol. 33, pp. 1-16, 2002.

[14]   Q. Song, "Seasonal forecasting in fuzzy time series," *Fuzzy Set Syst*, vol. 107, pp. 235-236, 1999.

[15]   E. Eğrioğlu, C.H. Aladag, U. Yolcu, M.A. Basaran, and V.R. Uslu, "A new hybrid approach based on SARIMA and partial high order bivariate fuzzy," *Expert Syst Appl*, vol. 36, pp. 7424-7434, 2009.

[16]   K. Huarng, "Heuristic models of fuzzy time series for forecasting," *Fuzzy Set Syst*, vol. 123, no. 3, 369-386, 2001.

[17]   Y.Y. Hsu, S.M. Tse, and B. Wu, "A new approach of bivariate fuzzy time series analysis to the forecasting of a stock index," *Int J Uncertain Fuzz*, vol. 11, no. 6, pp. 671-690, 2003.

[18]   T.K. Yu, and K. Huarng, "A bivariate fuzzy time series model to forecast the TAIEX," *Expert Syst Appl*, vol. 34, no. 4, pp. 2945-2952, 2008.

[19]   G.E.P. Box, and G.M. Jenkins, *Time Series Analysis: Forecasting and Control*. Holdan-Day: San Francisco, 1976.

[20]   J.M., Zurada, *Introduction of Artificial Neural Systems*. West Publishing: St. Paul, 1992.

[21]   G. Zhang, B.E. Patuwo, and Y.M. Hu, "Forecasting with artificial neural networks: the state of the art," *Int J Forecasting*, vol. 14, pp. 35-62, 1998.

[22]   P.J. Werbos, "*Beyond Regression: New tools for Prediction and Analysis in the Behavioral Sciences*" Ph. D. thesis, The Harvard University, Cambridge, 1974.

[23]   D.B. Parker, "Learning-Logic", Invention Report, Office of Technology Licensing Stanford University, pp. 81-64, file 1, Stanford, 1982.

[24]   D.E. Rumelhart, G.E. Hinton, and R.J. Williams, "Learning Representations by Backpropagating Errors," *Nature*, vol. 323, no. 6188, pp. 533-536, 1986.

[25]   Y.L. Chun, Ed., *A theoretical framework for backpropagation*. Morgan Kaufmann: Connectionist, 1988.

[26]   K.A. Smith, J.N.D. Gupta, *Neural networks in business: techniques and applications*. Idea Group: Hershey, 2002.

[27]   K. Levenberg, "A method for the solution of certain non-linear problems in least squares," *Q Appl Math*, vol. 2, pp. 164–168, 1944.

[28]   S.H. Ngia, "Efficient Training of Neural Nets for Nonlinear Adaptive Filtering Using a Recursive Levenberg-Marquardt Algorithm," *IEEE Trans on Signal Process*, vol. 48, pp. 1915-1927, 2000.

# CHAPTER 13

# Forecasting Gold Prices Series in Turkey by the Forecast Combination

## Cagdas Hakan Aladag[1,*], Erol Eğrioğlu[2] and Cem Kadilar[1]

[1]*Hacettepe University, Faculty of Science, Department of Statistics, 06800, Ankara, Turkey and* [2]*Ondokuz Mayis University, Faculty of Arts and Science, Department of Statistics, 55139, Samsun, Turkey*

**Abstract:** Forecast combination is a method used for obtaining more accurate forecasts. Forecast combination consists of the combination of forecasts obtained from different models with various methods. There are several types of forecast combination in the forecasting literature. In this study, various fuzzy time series approaches are applied to Turkey's daily highest gold prices series and forecasts obtained from these approaches are combined with variance covariance method (VCM), mean square forecast error method (MSFE) and artificial neural networks (ANN) approach. Results obtained from all of these methods are analyzed and the optimal forecast technique for Turkey's daily highest gold prices series is determined.

**Keywords:** Artificial neural networks, Forecasting, Forecast combination, Fuzzy time series.

## 1. INTRODUCTION

Numerous studies have been done and various approaches have been proposed in the literature to get better forecasts in time series. Among these methods, the most widely used two approaches in recent years are ANN and fuzzy time series methods which give very promising results for the forecasting problem [1, 2]. Another effective method used for the time series forecasting problem is the combination of forecasts of various methods. This method was first proposed in [3]. Then, [4-6] proposed forecast combination for more than two models. [7] compared three different forecast combinations for the combination of four time series models.

When forecast combination method is used, the forecasting value is obtained by a linear combination in which various models are combined with weights. The contribution of each model in forecasts of combination approach differs. The weights in forecast combination can be determined by both the particular assumptions and various optimization methods. The key point in forecast combination is the determination of optimal weights and combination function that give the best forecasts. Forecasts of combination function may be linear or curvilinear. [8] proposed a new forecast combination model in which forecasts obtained from fuzzy time series forecasting models are used as feed forward artificial neural network input and output of artificial neural network is obtained as combined forecasts. When this artificial neural network is optimized, optimal weights that provide the best curvilinear match up are determined. Therefore, when artificial neural networks method is used in forecast combination, both combination function and optimal weights can be determined without any problem.

In this study, various fuzzy time series approaches are applied to Turkey's daily highest gold prices series and forecasts obtained from the approaches are combined with variance covariance method (VCM), mean squared forecast error method (MSFE), and ANN approach. Results obtained from all of these methods are analyzed and the optimal forecasting technique for Turkey's daily highest gold prices series is developed.

In the second section of the study, forecast combination techniques existing in the literature are introduced. Third section deals with the general definition of fuzzy time series and a brief summary of fuzzy time series methods. In the fourth section, forecast combination techniques are applied to Turkey's daily highest gold prices series. Obtained results are discussed in the last section.

*Address correspondence to Cagdas Hakan Aladag: Hacettepe University, Faculty of Science, Department of Statistics, 06800, Ankara, Turkey; E-mail: chaladag@gmail.com

## 2. FORECAST COMBINATION TECHNIQUES

In the literature, various forecast combinations have been developed. Three of the best known are; simple forecast combination, variance covariance combination and mean squared forecast error method. [8] also proposed a forecast combination based on artificial neural networks.

### 2.1. Simple Forecast Combination (SFC) Method

In the simple forecast combination method, combined forecasts are obtained by multiplying and adding forecasts obtained from two or more models with weights. Combine forecast is calculated by using the formula given below.

$$f_c = \sum_{i=1}^{n} w_i f_i \tag{1}$$

where, $f_i$ and $n$ represent the forecast value obtained from the forecasting model $i$, and number of forecasting models, respectively. In this method, weights are not necessarily equal. If intervals are equal,

$$w_i = 1/n \tag{2}$$

then the method is called simple mean combination. As a result, the combined forecast, no matter what the weights are, is a linear combination of model forecasts.

### 2.2. Variance Covariance Method (VCM)

Variance covariance method is a linear forecast combination approach in which the weights are determined based on performance of model forecasts. In this method, weights are calculated by using the formula given below.

$$w' = u'\Sigma^{-1}/u'\Sigma^{-1}u \tag{3}$$

where $\Sigma$ represents covariance matrix, $u = (1,1,.,1)'$ and $\sum_{i=1}^{n} w_i = 1$. In some cases, weights may be negative and in this case, the use of variance covariance method is not suitable. When the weights are obtained according to equation (3), combined forecasts are calculated according to equation (1).

### 2.3. Mean Squared Forecast Error Method (MSFE)

Unlike variance-covariance method, in the MSFE method, combination weights are determined by increasing contribution of the last forecast. In this method, weights are calculated by using the formula as

$$w_i = \frac{1/\sum_{t=1}^{T} \beta^{T-t+1} e_{it}^2}{1/\sum_{i=1}^{n} \sum_{t=1}^{T} \beta^{T-t+1} e_{it}^2} \tag{4}$$

where $0 < \beta < 1$ is the arbitrary discount factor and $e_{it}^2$ is the forecast error obtained from model $i$ for observation $t$. When the weights are obtained according to equation (4), combined forecasts are calculated according to equation (1).

### 2.4. Forecast Combination Approach Based On Artificial Neural Network (ANNFC)

Although the first three combinations use different techniques in determining weights, combination functions of three methods are linear. Selection of linear combination function makes easy to obtain optimal weight values. However, curvilinear function is preferred in obtaining optimal combination. In [8], a forecast combination, in which forecasts obtained from various models are taken as input of artificial neural network and output of artificial neural network is obtained as combined forecast, was proposed. In ANN techniques proposed by [8], weights of combination function was a curvilinear function.

In ANNFC technique, the number of input units in artificial neural network used for the combination is equal to the number of forecasting models ($n$). Hidden layer unit number is taken as 1 to ensure the simple combination function structure. Output number is also taken as 1. Artificial neural network used for combination is given in Fig. **1**. Artificial neural network whose architecture, given in Fig. **1**, can be combined with forecasts obtained from three models.

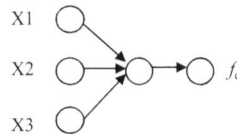

**Figure 1:** Feed forward ANN model for forecast combination.

Feed forward ANN model is used for forecast combination, logistic activation function, given in (5), is used for hidden layer units and a linear activation function, given in (6), is used for output layer.

$$f(net) = \frac{1}{1+\exp{(-net)}} \tag{5}$$

$$f(net) = net \tag{6}$$

Mathematical representation of feed forward artificial neural network can be given by

$$f_c = \left(\left(\sum_{i=1}^{3} w(1,i)\frac{1}{1+\exp(-X_i)}\right) + w(3,1)\right)w(2,1) + w(3,2) \tag{7}$$

where $x_i$ is the forecast obtained from model $i$, $w(1,i)$ is connection weights between input layer and hidden layer, where $i$=1,2,3, $w(3,1)$ is a connection weight of input layer side to weight hidden layer, $w(2,1)$ is a connection weight between hidden layer and output layer, and $w(3,2)$ is a weight between hidden layer side and output layer. It is clear that function, given in (7), is in a curvilinear form. The weights obtained from this function are optimal weights for forecast combination and output of artificial neural network is combined forecasts.

## 3. FUZZY TIME SERIES METHODS

Fuzzy time series approaches depend on basic definitions defined as follows: Let $U = \{u_1,...,u_b\}$ be the universe of discourse, where $u_1$, ..., $u_b$ are intervals. These intervals are obtained from partition of the universe of discourse including all values of time series. A fuzzy set $A_i$ of $U$ is defined as

$$A_i = f_{A_i}(u_1)/u_1 + \cdots + f_{A_i}(u_b)/u_b$$

where $f_{A_i}$ is the membership function of the fuzzy set $A_i$, $f_{A_i}:U \to [0,1]$. $f_{A_i}(u_b)$ is the degree of belongingness of $u_b$ to $A_i$.

**Definition 1.** Let $Y(t)$, $t =...,0,1,2,...$ be a subset of real numbers. After defining the universe of discourse, the new time series $F(t)$ consisting of $A_i$ is called a fuzzy time series.

**Definition 2.** Fuzzy time series relationships assume that $F(t)$ is caused only by $F(t-1)$. Then, the fuzzy relationship is called the first order fuzzy time series. The first order fuzzy time series forecasting model can be written as follows:

$$F(t-1) \to F(t) \tag{8}$$

**Definition 3.** Let $F(t)$ be a fuzzy time series. If $F(t)$ is caused by $F(t-1), F(t-2),..., $ and $F(t-n)$, then this fuzzy logical relationship is represented by

$$F(t-n),...,F(t-2),F(t-1) \rightarrow F(t) \qquad\qquad (9)$$

and it is called the n[th] order fuzzy time series forecasting model.

The definition of fuzzy time series approach was first proposed in [9-11], the idea of utilization of fuzzy time series table was presented. In [12], this approach was extended to high order models. [2] proposed a fuzzy time series approach of forecasting based on feed forward artificial neural networks for high order models. While [11] was using an approach based on a first order model, [12] and [8] were utilizing an approach based on a high order model.

## 4. APPLICATION

Turkey's daily highest gold prices (TL/kg) series (HGP) can be obtained from the website of Central Bank of The Turkish Republic between 06 October 2008 and 30 April 2009. In this section, time series analysis for HGP is given in detail. HGP has some missing observations; therefore, these observations are imputed by the mean of nearby method. Graph of the time series, after imputing the missing data, is given in Fig. **2**.

**Figure 2:** Time series graph of HGP.

HGP has 233 observations. In this section, application is given in two parts. In the first part, fuzzy time series approaches, suggested by [2, 11, 12], are used and in the second part, the method of the combination of VCM, MSFE and ANN is used for the time series analysis of HGP.

### 4.1. Fuzzy Time Series Methods

As definition of the universe of discourse includes all of three time series methods, the universe of discourse is defined for HGP as $U$= [30600, 54600]. Since all of three fuzzy time series methods are based on universe of discourse partition, interval lengths enabling partition of universe of discourse are selected as 7, 8, 9, 10, and 11, respectively. These interval lengths are 3420, 2980, 2650, 2387, and 2170, respectively. In cases where the length of interval is 3420, sub intervals are as follows:

$$u_1 = [30600, 34020], u_2 = [34020, 37440], u_3 = [37440, 40860], u_4 = [40860, 44280], u_5 = [44280, 47700], u_6 = [47700, 51120], u_7 = [51120, 54600]$$

By these interval lengths, the method suggested by [11] is first applied to HGP for fuzzy time series methods. Two hundred observations of HGP during the period of October 10, 2008 and March 16, 2009 are used as training data, whereas 33 observations during the period of March, 17 2009 and April, 30 2009 are used as test data. Root mean square error (RMSE) values obtained from this method to HGP for the test data, are given in Table **1**.

**Table 1:** Results of the method by [11] for HGP test data

| *Length of Intervals* | | | | |
|---|---|---|---|---|
| 3420 | 2980 | 2650 | 2387 | 2170 |
| **RMSE** 1652 | 1256 | 1525 | 1471 | 1626 |

In Table **1**, the lowest RMSE value is 1256 for the case where the universe of discourse is partitioned when the interval length is 2980.

Secondly, high order fuzzy time series proposed by [12] is applied to the data. Model order is taken from 2 to 7 and 30 different analyses are done. RMSE values for different model orders and interval lengths are shown in Table **2**.

**Table 2:** Results of the method by [12] for HGP test data

| *Length of Intervals* | *Model Order* | | | | | |
|---|---|---|---|---|---|---|
| | *2* | *3* | *4* | *5* | *6* | *7* |
| 3420 | 2785.081 | 2785.081 | 2650.112 | 1928.49 | 1925.41373 | 1953.461 |
| 2980 | 2309.725 | 2290.842 | 1680.441 | 1634.451 | 1671.65621 | 1722.488 |
| 2650 | 2247.682 | 2300.11 | 1523.527 | 1397.54 | 1460.31585 | 1479.701 |
| 2387 | 1953.984 | 1923.282 | 1818.526 | 1392.613 | 1389.52438 | 1458.993 |
| 2170 | 1567.093 | 1606.067 | 1500.834 | 1511.719 | 1341.2199 | 1361.177 |

In Table **2**, it is observed that the lowest RMSE value (1341.2199).is obtained when the model order is 6 and the length of interval is 2170.

The third fuzzy time series method used in the analysis of HGP is the method suggested by [2] which is a high order model and consists of feed forward artificial neural networks. Again, the model order is taken from 2 to 7, the number of feed forward artificial neural network for hidden layer is taken from 1 to 4 and interval lengths are taken as 3420, 2980, 2650, 2387, and 2170. RMSE values for this method are shown in Tables **3-8**, respectively.

**Table 3:** Results of the method by [2] for HGP test data when the model order is 2

| *Length Of Intervals* | *Number of Hidden Layer Neurons* | | | |
|---|---|---|---|---|
| | *1* | *2* | *3* | *4* |
| 2170 | 1444.07 | 1444.07 | 1286.99 | 1574.89 |
| 2387 | 1351.44 | 1351.44 | 1188.20 | 1188.20 |
| 2650 | 1362.76 | 1362.76 | 1367.46 | 1243.52 |
| 2980 | 1407.38 | 1579.79 | 1256.02 | 1407.38 |
| 3420 | 1144.63 | 1144.63 | 1318.79 | 1318.80 |

**Table 4:** Results of the method by [2] for HGP test data when the model order is 3

| *Length Of Intervals* | *Number of Hidden Layer Neurons* | | | |
|---|---|---|---|---|
| | *1* | *2* | *3* | *4* |
| 2170 | 1469.35 | 1469.35 | 1406.61 | 1730.65 |
| 2387 | 1360.84 | 1530.98 | 1368.95 | 1188.20 |
| 2650 | 1417.63 | 1417.63 | 1409.68 | 1192.76 |
| 2980 | 1407.38 | 1384.09 | 1256.03 | 2153.72 |
| 3420 | 1144.63 | 1938.57 | 1938.57 | 1938.57 |

**Table 5:** Results of the method by [2] for HGP test data when the model order is 4

| Length Of Intervals | Number of Hidden Layer Neurons | | | |
|---|---|---|---|---|
| | 1 | 2 | 3 | 4 |
| 2170 | 1657.76 | 1657.76 | 1531.07 | 1373.26 |
| 2387 | 1582.05 | 1582.05 | 1723.46 | 1756.05 |
| 2650 | 1437.32 | 1437.32 | 1409.68 | 1047.59 |
| 2980 | 1407.38 | 1407.38 | 1488.76 | 1812.12 |
| 3420 | 1406.63 | 1406.63 | 1938.57 | 1839.26 |

**Table 6:** Results of the method by [2] for HGP test data when the model order is 5

| Length Of Intervals | Number of Hidden Layer Neurons | | | |
|---|---|---|---|---|
| | 1 | 2 | 3 | 4 |
| 2170 | 1770.99 | 1442.84 | 1532.79 | 1423.11 |
| 2387 | 1996.07 | 1604.71 | 2544.04 | 1479.67 |
| 2650 | 1521.46 | 1475.92 | 1689.52 | 1626.56 |
| 2980 | 1683.03 | 1546.29 | 3064.86 | 1546.29 |
| 3420 | 1730.40 | 1911.39 | 2006.09 | 1837.58 |

**Table 7:** Results of the method by [2] for HGP test data when the model order is 6

| Length Of Intervals | Number of Hidden Layer Neurons | | | |
|---|---|---|---|---|
| | 1 | 2 | 3 | 4 |
| 2170 | 1302.74 | 1491.25 | 1583.22 | 2296.70 |
| 2387 | 1315.66 | 1488.05 | 8534.19 | 1558.24 |
| 2650 | 1403.40 | 1233.15 | 1585.31 | 2024.84 |
| 2980 | 1407.38 | 1663.22 | 1550.85 | 1762.34 |
| 3420 | 1144.63 | 1571.91 | 1987.15 | 1573.23 |

**Table 8:** Results of the method by [2] for HGP test data when the model order is 7

| Length Of Intervals | Number of Hidden Layer Neurons | | | |
|---|---|---|---|---|
| | 1 | 2 | 3 | 4 |
| 2170 | 1299.96 | 8129.78 | 5543.26 | 3547.20 |
| 2387 | 1391.49 | 8234.26 | 4256.69 | 6615.03 |
| 2650 | 1403.40 | 8341.73 | 3935.55 | 1884.97 |
| 2980 | 1407.38 | 8491.30 | 7001.82 | 3348.86 |
| 3420 | 1144.63 | 2782.79 | 4234.30 | 1962.48 |

In Tables **3-8** the lowest RMSE values are shown in bold. From these tables, it is seen that the lowest RMSE value is 1047. Therefore, it is inferred that the best forecasts are obtained when the order of feed forward artificial neural network is 4, the interval length is 2650, and the number of hidden layer is 4.

The graphs of best forecasts of the methods in [2, 11, 12], and [2] for the test data, are given in Fig. **3** and corresponding values are shown in Table **9**.

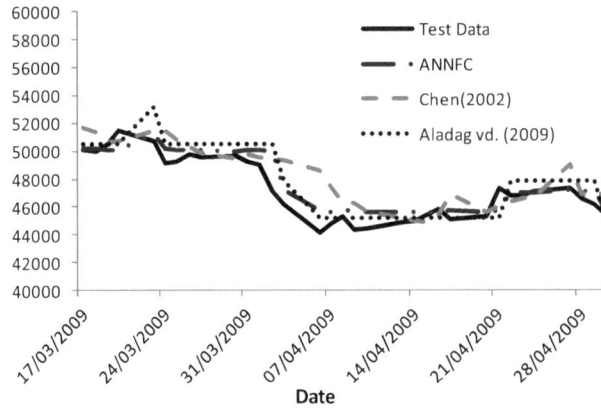

**Figure 3:** Forecasts of Fuzzy Time Series Methods and HGP Test Data.

**Table 9:** HGP Test Data and Best Forecasts of Methods in [2, 11, 12]

| Date | HGP Test Data | [11] | [12] | [2] |
|------|---------------|------|------|-----|
| 17.03.2009 | 50050 | 49970 | 51628.33 | 50475 |
| 18.03.2009 | 49970 | 49970 | 51318.33 | 50475 |
| 19.03.2009 | 50530 | 49970 | 50595.00 | 50475 |
| 20.03.2009 | 51430 | 49970 | 50698.33 | 50475 |
| 23.03.2009 | 50740 | 52950 | 51421.66 | 53125 |
| 24.03.2009 | 49070 | 49970 | 51421.66 | 50475 |
| 25.03.2009 | 49220 | 49970 | 50801.66 | 50475 |
| 26.03.2009 | 49710 | 49970 | 50285.00 | 50475 |
| 27.03.2009 | 49550 | 49970 | 49871.66 | 50475 |
| 30.03.2009 | 49600 | 49970 | 49458.33 | 50475 |
| 31.03.2009 | 49260 | 49970 | 49768.33 | 50475 |
| 01.04.2009 | 49000 | 49970 | 49561.66 | 50475 |
| 02.04.2009 | 47100 | 49970 | 49458.33 | 50475 |
| 03.04.2009 | 46170 | 46990 | 49355.00 | 47825 |
| 06.04.2009 | 44100 | 46990 | 48631.66 | 45175 |
| 07.04.2009 | 44800 | 44010 | 47391.66 | 45175 |
| 08.04.2009 | 45300 | 46990 | 46358.33 | 45175 |
| 09.04.2009 | 44345 | 46990 | 46255.00 | 45175 |
| 10.04.2009 | 44500 | 44010 | 45635.00 | 45175 |
| 13.04.2009 | 44860 | 44010 | 45221.66 | 45175 |
| 14.04.2009 | 45000 | 44010 | 45015.00 | 45175 |
| 15.04.2009 | 45400 | 44010 | 44911.66 | 45175 |
| 16.04.2009 | 45800 | 46990 | 45428.33 | 45175 |
| 17.04.2009 | 45110 | 46990 | 46875.00 | 45175 |
| 20.04.2009 | 45255 | 44010 | 45635.00 | 45175 |
| 21.04.2009 | 47300 | 46990 | 46048.33 | 45175 |
| 22.04.2009 | 46734 | 46990 | 46358.33 | 47825 |
| 23.04.2009 | 46734 | 46990 | 46565.00 | 47825 |
| 24.04.2009 | 47090 | 46990 | 46668.33 | 47825 |
| 27.04.2009 | 47290 | 46990 | 49045.00 | 47825 |
| 28.04.2009 | 46620 | 46990 | 46875.00 | 47825 |
| 29.04.2009 | 46180 | 46990 | 46875.00 | 47825 |

*Table 9: cont....*

| | | | | |
|---|---|---|---|---|
| 30.04.2009 | 45365 | 46990 | 46875.00 | 45175 |
| | **RMSE** | 1256 | 1341.22 | 1047.59 |

## 4.2. Forecast Combination Methods

Forecasts obtained from implementing three different fuzzy time series approaches in HGP are combined using forecast combination techniques. Variance covariance method is first applied. When variance covariance method is applied for the forecasts obtained from methods by [2, 11, 12], one of the weights is found as negative. Therefore, in the application of VCM, forecast obtained from methods by [2, 11, 12] are combined. Optimal weights obtained from the forecast combination method are obtained as $w_1 = 0.7394$ and $w_2 = 0.2605$. Combined forecasts for this method are shown in Table **11**. At the end of the combination, the RMSE value for the VCM method is found as 1158.10.

Secondly, MSFE forecast combination method is applied to the data. Forecasts, which are obtained from fuzzy time series methods in [2, 11, 12] and shown in Table 9, are combined. In the implementation of MSFE method, discount factor is arbitrary taken as $\beta = 0.8$. Optimal weights for the MSFE method are obtained as $w_1 = 0.3245, w_2 = 0.2779$ and $w_3 = 0.3976$. Combined forecasts for this method are shown in Table 11. At the end of the combination, the RMSE value for the MSFE method is obtained as 1001.76. When it is compared with the RMSE of VCM, forecasting with MSFE has lower RMSE value.

Finally, forecasts are combined using the ANNFC method. In the application of the method, forecasts, which are obtained from methods by [2, 11, 12] and shown in Table **9**, are combined. Optimal ANN weights for the ANNFC method are shown in Table **10**.

**Table 10:** Optimal weights of feed forward ANN in ANNFC Method

| Weights | Optimal Values |
|---|---|
| w(1,1) | 0.98598206 |
| w(1,2) | 0.38766465 |
| w(1,3) | 4.91135656 |
| w(3,1) | -2.9682975 |
| w(2,1) | 0.73420973 |
| w(3,2) | 0.09383837 |

Forecasts obtained from the ANNFC method are given in Table **11** and RMSE value for this method is found as 847.11 which is the lowest value according to MSE values of VCM and MSFE methods.

**Table 11:** HGP Test Data and Forecast Results of VCM, MSFE, ANNFC Methods

| Date | HGP Test Data | VCM | MSFE | ANFC |
|---|---|---|---|---|
| 17.03.2009 | 50050 | 51261.87 | 50631.66 | 50144.94 |
| 18.03.2009 | 49970 | 50998.30 | 50545.51 | 50126.86 |
| 19.03.2009 | 50530 | 50383.30 | 50344.49 | 50083.96 |
| 20.03.2009 | 51430 | 50471.15 | 50373.20 | 50090.15 |
| 23.03.2009 | 50740 | 51609.64 | 52594.85 | 51192.18 |
| 24.03.2009 | 49070 | 51086.16 | 50574.22 | 50132.91 |
| 25.03.2009 | 49220 | 50559.01 | 50401.92 | 50096.32 |
| 26.03.2009 | 49710 | 50119.73 | 50258.33 | 50065.26 |
| 27.03.2009 | 49550 | 49768.30 | 50143.47 | 50040.03 |
| 30.03.2009 | 49600 | 49416.87 | 50028.60 | 50014.46 |
| 31.03.2009 | 49260 | 49680.44 | 50114.75 | 50033.67 |
| 01.04.2009 | 49000 | 49504.72 | 50057.32 | 50020.88 |

*Table 11: cont….*

| | | | | |
|---|---|---|---|---|
| 02.04.2009 | 47100 | 49416.87 | 50028.60 | 50014.46 |
| 03.04.2009 | 46170 | 48805.53 | 47979.26 | 47238.91 |
| 06.04.2009 | 44100 | 48810.25 | 46724.57 | 45760.76 |
| 07.04.2009 | 44800 | 46612.76 | 45413.01 | 45605.12 |
| 08.04.2009 | 45300 | 46877.39 | 46092.79 | 45703.72 |
| 09.04.2009 | 44345 | 46789.53 | 46064.07 | 45701.27 |
| 10.04.2009 | 44500 | 45119.18 | 44924.82 | 45573.61 |
| 13.04.2009 | 44860 | 44767.76 | 44809.95 | 45566.61 |
| 14.04.2009 | 45000 | 44592.04 | 44752.51 | 45563.16 |
| 15.04.2009 | 45400 | 44504.18 | 44723.80 | 45561.45 |
| 16.04.2009 | 45800 | 46086.67 | 45834.34 | 45682.16 |
| 17.04.2009 | 45110 | 47316.67 | 46236.38 | 45716.13 |
| 20.04.2009 | 45255 | 45119.18 | 44924.82 | 45573.61 |
| 21.04.2009 | 47300 | 46613.81 | 46006.64 | 45696.42 |
| 22.04.2009 | 46734 | 46257.67 | 47146.46 | 47014.33 |
| 23.04.2009 | 46734 | 46433.38 | 47203.90 | 47029.32 |
| 24.04.2009 | 47090 | 46521.24 | 47232.61 | 47036.84 |
| 27.04.2009 | 47290 | 48541.96 | 47893.11 | 47214.98 |
| 28.04.2009 | 46620 | 46696.95 | 47290.05 | 47051.95 |
| 29.04.2009 | 46180 | 46696.95 | 47290.05 | 47051.95 |
| 30.04.2009 | 45365 | 47316.67 | 46236.38 | 45716.13 |
| **RMSE** | | 1158.11 | 1001.77 | 847.12 |

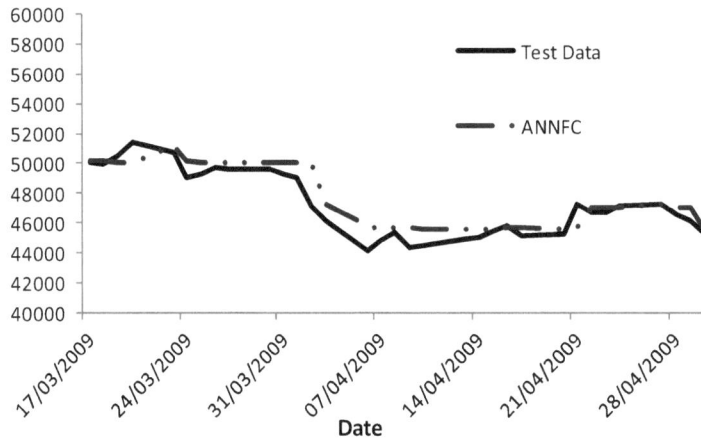

**Figure 4:** Forecasts of ANNFC Method and HGP Test Data.

## 5. CONCLUSION

In this study, Turkey's daily highest gold prices series is analyzed by various fuzzy time series methods such as the methods in [2, 11, 12] that are the best known methods in literature. The method in [11] is a first order model, whereas the methods in [2, 12] have high order models. While [11, 12] determine fuzzy relationships based on fuzzy group relation tables, [2] use feed forward artificial neural networks in the analysis of fuzzy relationships. Owing to the fact that all of three methods have different feature, they give different forecast results. In this study, forecasts obtained from these different fuzzy time series methods are combined with VCM, MSFE, and ANNFC techniques and more accurate results are obtained. The implementation of fuzzy time series merely gives the best results in the method of [2] and RMSE value is found as 1047.59. When forecast combination techniques are applied, forecasts become more accurate as RMSE value is obtained as 1001.77 in MSFE method but this decreases to 847.12 in ANNFC method. It shows that the combination of forecasts obtained from methods in [2, 11, 12] with ANNFC method gives the best forecasts for HGP. The graph of the forecasts by ANNFC method with the real data is shown in Fig. **4**.

## REFERENCES

[1]     E. Eğrioğlu, C.H. Aladag, and S. Gunay, "A new model selection strategy in artificial neural network" *Appl Math Comput*, vol. 195, pp. 591–597, 2008.

[2]     C.H. Aladag, M.A. Basaran, E. Eğrioğlu, U. Yolcu, and V.R. Uslu, "Forecasting in high order fuzzy times series by using neural networks to define fuzzy relations," *Expert Syst Appl*, vol. 36, no. 3, pp. 4228–4231, 2009.

[3]     J.M. Bates, C.W.J. Granger, "The Combination of forecast," *Oper Res Quart*, vol. 20, no. 4, pp. 451-468, 1969.

[4]     C.W.J. Granger, and R. Ramanathan, "Improved methods of combined forecasts," *J Forecasting*, vol. 3, pp. 197-204, 1984.

[5]     P. Newbold, and C.W.J. Granger, "Experience with forecasting time series and combination of forecasts," *J R Stat Soc A*, vol. 137, no. 2, pp. 131-165, 1974.

[6]     R.L. Winkler, and S. Markidakis, "The Combination of Forecasts," *J R Stat Soc A*, vol. 146 (2), pp. 150-157,1983.

[7]     K.K.F. Wong, H. Song, S.F. Witt, and D.C. Wu, "Tourism forecasting: to combine or not to combine?," *Tourism Manage*, vol. 28, pp. 1068-1078, 2007.

[8]     C.H. Aladag, E. Eğrioğlu, and U. Yolcu, "Forecast combination by using artificial neural networks," *Neural Process Lett*, vol. 32 (3), pp. 269–276, 2010.

[9]     Q. Song, and B.S. Chissom, "Fuzzy time series and its models," *Fuzzy Set Syst*, vol. 54, pp. 269-277, 1993.

[10]    Q. Song, and B.S. Chissom, "Forecasting enrollments with fuzzy time series - Part I," *Fuzzy Set Syst*, vol. 54, pp. 1-10, 1993.

[11]    S.M. Chen, "Forecasting enrollments based on fuzzy time-series," *Fuzzy Set Syst*, vol. 81, pp. 311-319, 1996.

[12]    S.M. Chen, "Forecasting enrollments based on high order fuzzy time series," *Cybernet Syst*, vol. 33, pp. 1-16, 2002.

# CHAPTER 14

# A Hybrid Forecasting Model Based on Multivariate Fuzzy Time Series and Artificial Neural Networks

## Cagdas Hakan Aladag[1,*] and Erol Eğrioğlu[2]

[1]*Hacettepe University, Faculty of Science, Department of Statistics, 06800, Ankara, Turkey and* [2]*Ondokuz Mayis University, Faculty of Arts and Science, Department of Statistics, 55139, Samsun, Turkey*

**Abstract:** Fuzzy time series approaches have been recently used for forecasting in many studies [1]. These approaches can be categorized into two subclasses that are univariate and multivariate approaches. It is a fact that many factors can actually affect real time series data. Therefore, using a multivariate fuzzy time series forecasting model can be more reasonable in order to get more accurate forecasts. The most preferred method is using tables of fuzzy relations for determining fuzzy relations in multivariate fuzzy time series approaches in the literature. However, employing this method is a computationally though task. In this study, we propose a new method based on utilizing artificial neural networks in determining fuzzy logic relations and using the formula defined by Jilani and Burney [2] in calculating defuzzyfied forecasts. Hence, it is aimed to produce more accurate forecasts and avoid intense computations. The proposed method is applied to the time series data of the total number of annual car road accidents casualties in Belgium from 1974 to 2004 and a comparison is made between our proposed method and the methods proposed by Jilani and Burney [2] and Lee *et al.* [3].

**Keywords:** Artificial neural networks, Forecasting, Fuzzy time series, Multivariate fuzzy time series approaches.

## 1. INTRODUCTION

The real time series data such as temperature and share prices of stockholders contain some uncertainty in itself. The conventional time series analyses can produce unsatisfactory forecasts for such time series. Song and Chissom [4-6] first introduced the definition of a fuzzy time series based on the concept of the fuzzy set theory proposed by Zadeh [7]. The Song and Chissom's model [4] is a fuzzy time series-forecasting model called as a one factor first order model. Chen [8] and Aladag *et al.* [1] present approaches based on a one factor high order fuzzy time series forecasting model. Using a multivariate fuzzy time series forecasting model can give more accurate forecasts since the real time series data can be affected by many factors. In order to forecast such time series, two-factor fuzzy time series model was used by Yu and Huarng [9], and Lee *et al.* [3] in the literature. Then, Jilani and Burney [2] analyze $k$-factor ($k \geq 1$) and $n^{th}$ order fuzzy time series forecasting models.

To obtain forecasts for fuzzy time series, using multivariate fuzzy time series model, instead of using univariate one, can provide better forecasts since real time series data has a complex structure and is affected by many other factors. In the literature, Yu and Huarng [9] presented an algorithm that analyzes a first order bivariate fuzzy time series-forecasting model. Then, Yu and Huarng [9] proposed another algorithm that analyzes a two-factor and first order fuzzy time series forecasting model using a feed forward artificial neural network. On the other hand, Jilani and Burney [2] proposed an algorithm based on an approach that is used to forecast a multivariate high order fuzzy time series. It has been shown that the method proposed by Jilani and Burney [2] gives better-forecast values than those generated by the method introduced by Lee *et al.* [3]. However, the determination of fuzzy relationships in the algorithm proposed by Jilani and Burney [2] depends on fuzzy logic relation tables which require too many complicated calculations and so too much time.

*Address correspondence to Cagdas Hakan Aladag: Hacettepe University, Faculty of Science, Department of Statistics, 06800, Ankara, Turkey; E-mail: chaladag@gmail.com

In this chapter, we proposed a new method to analyze $k$-factor and $n^{th}$ order fuzzy time series-forecasting model. In the proposed method, feed forward artificial neural networks (ANN) are utilized to determine fuzzy logic relations. In addition, we use the formula defined by Jilani and Burney [2] in the defuzzyfication step. The proposed method is applied to the total number of annual car road accidents casualties in Belgium and the results obtained from the proposed method are compared with those obtained from the methods by Jilani and Burney [2] and Lee *et al.* [3]. We would like to note that the method proposed in this chapter is different from the method proposed by Eğrioğlu *et al.* [12] although some parts of this chapter were taken from the study of Eğrioğlu *et al.* [12].

Section 2 gives the fundamental fuzzy time series definition. The brief information related to ANN is given in Section 3. The proposed method is introduced in Section 4. Section 5 is the implementation of the proposed method using the data of the total number of annual car road accidents casualties in Belgium. Section 6 is the last part that gives comparison results and discussion.

## 2. FUZZY TIME SERIES

The definition of fuzzy time series was first introduced by Song and Chissom [4,5]. The definitions related to fuzzy time series are given as follows:

Let $U$ be the universe of discourse, where $U = \{u_1, u_2, ..., u_b\}$. A fuzzy set $A_i$ of $U$ is defined as $A_i = f_{A_i}(u_1)/u_1 + f_{A_i}(u_2)/u_2 + \cdots + f_{A_i}(u_b)/u_b$, where $f_{A_i}$ is the membership function of the fuzzy set $A_i; f_{A_i} : U \to [0,1]$. $u_a$ is a generic element of fuzzy set $A_i; f_{A_i}(u_a)$ is the degree of belongingness of $u_a$ to $A_i; f_{A_i}(u_a) \in [0,1]$ and $1 \le a \le b$.

***Definition 1.*** **Fuzzy time series.** Let $Y(t)$ $(t = ..., 0, 1, 2, ...)$, a subset of real numbers be the universe of discourse by which fuzzy sets $f_j(t)$ are defined. If $F(t)$ is a collection of $f_1(t), f_2(t),...$ then $F(t)$ is called a fuzzy time series defined on $Y(t)$.

***Definition 2.*** Fuzzy time series relationships assume that $F(t)$ is caused only by $F(t-1)$, then the relationship can be expressed as: $F(t) = F(t-1) * R(t, t-1)$, which is the fuzzy relationship between $F(t)$ and $F(t-1)$, where * represents as an operator. To sum up, let $F(t-1) = A_i$ and $F(t) = A_j$. The fuzzy logical relationship between $F(t)$ and $F(t-1)$ can be denoted as $A_i \to A_j$ where $A_i$ refers to the left-hand side and $A_j$ refers to the right-hand side of the fuzzy logical relationship. Furthermore, these fuzzy logical relationships can be grouped to establish different fuzzy relationships.

***Definition 3.*** Let $F(t)$ be a fuzzy time series. If $F(t)$ is caused by $F(t-1), F(t-2), ...,$ and $F(t-n)$, then this fuzzy logical relationship is represented by

$$F(t-n), ..., F(t-2), F(t-1) \to F(t) \tag{1}$$

and it is called the $n^{th}$ order fuzzy time series forecasting model.

***Definition 4.*** Let F and G be two fuzzy time series. Suppose that $F(t-1) = A_i$, $G(t-1) = B_k$ and $F(t) = A_j$. A bivariate fuzzy logical relationship is defined as $A_i, B_k \to A_j$, where $A_i, B_k$ are referred to as the left hand side and $A_j$ as the right hand side of the bivariate fuzzy logical relationship.

Therefore, first order bivariate fuzzy time series forecasting model is as follows:

$$F(t-1), G(t-1) \to F(t) \tag{2}$$

***Definition 5.*** Let F and G be two fuzzy time series. If $F(t)$ is caused by

$$(F(t-1), G(t-1)), (F(t-2), G(t-2)), ..., (F(t-n), G(t-n))$$

then this fuzzy logical relationship is represented by

$$(F(t-1), G(t-1)), (F(t-2), G(t-2)), \ldots, (F(t-n), G(t-n)) \rightarrow F(t) \qquad (3)$$

and it is called the two-factors $n^{\text{th}}$ order fuzzy time series forecasting model, where $F(t)$ and $G(t)$ are called the main factor fuzzy time series and the second factor fuzzy time series, respectively ($t=\ldots 0,1,2,\ldots$).

**Definition 6.** Let $F$ and $G_1, G_2 \ldots G_{k-1}$ be $k$ fuzzy time series. If $F(t)$ is caused by

$(F(t-1), G_1(t-1)G_2(t-1)\ldots G_{k-1}(t-1)), \ldots, (F(t-n), G_1(t-n)G_2(t-n)\ldots G_{k-1}(t-n))$    then    this    fuzzy logical relationship is represented by

$$(F(t-1), G_1(t-1)G_2(t-1)\ldots G_{k-1}(t-1)), \ldots,$$

$$(F(t-n), G_1(t-n)G_2(t-n)\ldots G_{k-1}(t-n)) \rightarrow F(t) \qquad (4)$$

and it is called the $k$-factors $n^{\text{th}}$ order fuzzy time series forecasting model, where $F(t)$ and $G_i(t)$ are called the main factor fuzzy time series and the secondary factors fuzzy time series, respectively $(t = \ldots 0,1,2,\ldots, i = 1,2,\ldots k-1)$.

## 3. ARTIFICIAL NEURAL NETWORKS

ANN consists of algorithms that mimic the features of brain of human being. These features are generating new knowledge and exploring by learning. In other words, ANN are synthetic networks that imitate biological neural networks. ANN are much more different than biological ones in terms of structure and ability [10]. ANN compose of a mathematical model [11]. The fundamental elements of ANN can be given as follows:

*Architecture structure*: The structure of multilayer feed forward ANN is basically given in Fig. **1**. Multilayer feed forward ANN as illustrated in the figure consist of three parts such as input, hidden, and output layers. Each layer consists of neurons. The architecture structure is determined based on deciding the number of neuron in each layer. These neurons are linked each other by weights. There is no link among the neurons in the same layer.

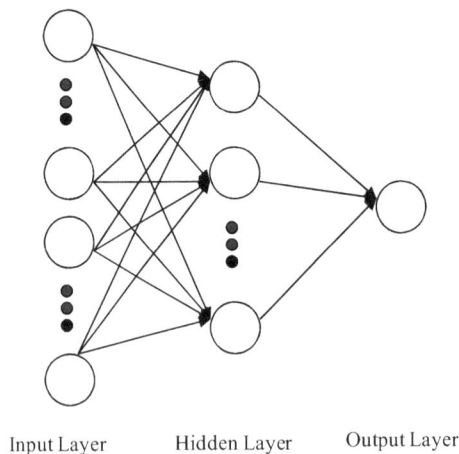

Input Layer          Hidden Layer          Output Layer

**Figure 1:** Multilayer feed forward ANN with one output neuron.

*Learning algorithm*: There have been many learning algorithms in order to determine weights. The one of the most employed algorithm is called Back Propagation Learning Algorithm. This learning algorithm updates the weights based on difference between real value and output value of the ANN. Learning parameter in back propagation algorithm plays an important role in reaching the desired outputs. When learning parameter is taken as a fixed value, it can be dynamically updated in the algorithm.

*Activation function*: Activation function provides the non-linear mapping between input and output. The performance of networks depends on the proper choice of activation function. Activation function can be chosen as either linear or double polarized, or one polarized. Slope parameter should be determined when the activation is non-linear. Also, slope parameter plays a key role in reaching desired output values.

## 4. THE PROPOSED METHOD

In this chapter, a new algorithm that determines fuzzy relationships using a feed forward artificial neural network and defuzzyfies fuzzy forecasts employing the formula defined by Jilani and Burney [2] is proposed in order to improve the accuracy of forecast for a multi-factor high order fuzzy time series. The proposed algorithm is given as follows.

**Step 1.** The universe of discourses and subintervals are defined for two time series. The min and max values of the time series are denoted by $D_{\min}$ and $D_{\max}$, respectively. Then two arbitrary positive numbers $D_1$ and $D_2$ for each fuzzy time series can be chosen in order to define the universe of discourse $U = [D_{\min} - D_1, D_{\max} + D_2]$.

**Step 2.** Define fuzzy sets based on the universe of discourses. Based on defined $U$ and $V$ universe of discourses and subintervals, fuzzy sets $A_1, A_2, ..., A_{r_1}$ and $B_{i1}, B_{i2}, ..., B_{ir_2}$ are defined and given below for time series and residuals, respectively.

$$A_1 = a_{11} / u_1 + a_{12} / u_2 + ... + a_{1r_1} / u_{r_1}$$
$$A_2 = a_{21} / u_1 + a_{22} / u_2 + ... + a_{2r_1} / u_{r_1}$$
$$\vdots$$
$$A_{r_1} = a_{r_1 1} / u_1 + a_{r_1 2} / u_2 + ... + a_{r_1 r_1} / u_{r_1}$$

$$B_{i1} = b_{i11} / v_{i,1} + b_{i12} / v_{i,2} + ... + b_{i1r_2} / v_{i,r_2}$$
$$B_{i2} = b_{i21} / v_{i,1} + b_{i22} / v_{i,2} + ... + b_{i2r_2} / v_{i,r_2}$$
$$\vdots$$
$$B_{ir_2} = b_{ir_2 1} / v_{i,1} + b_{ir_2 2} / v_{i,2} + ... + b_{ir_2 r_2} / v_{i,r_2}$$

where $a_{ij}$ is the degree of membership values of $u_i$ and $a_{ij} \in [0,1]$ for $1 \le i, j \le r_1$. $r_1$ is the numbers of subintervals and fuzzy sets, respectively. In similar way, where $b_{ijm}$ is the degree of membership values of $v_{ij}$ and $b_{ijm} \in [0,1]$ for $1 \le i \le k-1$ and $1 \le j, m \le r_2$. $r_2$ is the numbers of subintervals and fuzzy sets, respectively.

$$A_j = a_{j1} / u_1 + a_{j2} / u_2 + \cdots + a_{jr_1} / u_{r_1},$$

$$a_{jm} = \begin{cases} 1 & ,m = j \\ 0.5 & ,m = j-1 \\ 0.5 & ,m = j+1 \\ 0 & ,o.w. \end{cases} , j,m=1,...,r_1$$

and

$$B_{ij} = b_{ij1} / v_{i,1} + b_{ij2} / v_{i,2} + \cdots + b_{ij,r_2} / v_{i,r_2} \, , \, i=1,2,\ldots,k\text{-}1$$

$$b_{ijm} = \begin{cases} 1 & ,m = j \\ 0.5 & ,m = j-1 \\ 0.5 & ,m = j+1 \\ 0 & ,o.w. \end{cases} ,j,m=1,\ldots,r_2$$

**Step 3.** Fuzzify the observations. Each crisp value is mapped into a fuzzy set where its membership degree has maximum value. Fuzzy main factor time series is demonstrated by $F(t)$, fuzzy secondary factors time series are demonstrated by $G_i(t)$, $i=1,2,\ldots k\text{-}1$.

**Step 4.** Establish fuzzy relationship. In order to establish fuzzy relationships, ANN is employed. The lagged variables $F(t-1),F(t-2),\ldots,F(t-n)$ of the main factor fuzzy time series are used for training. and the lagged variables $G_1(t\text{–}1)$, $G_1(t\text{–}2)$, $\ldots$, $G_1(t\text{–}n)$, $\ldots$, $G_{k-1}(t\text{–}1)$, $G_{k-1}(t\text{–}2)$, $\ldots$, $G_{k-1}(t\text{–}1)$ of secondary factors are taken as the inputs of network. The main factor fuzzy time series $F(t)$ is used for the output of network. Feed forward neural network is trained in terms of these inputs and output. The number of neurons in the input layer is taken as the product of $k$ times $n$ while the number of neurons in the hidden layer (NNHL) can be decided by trial and error method. It is obvious that the number of neurons in the output layer should be one.

A small example will give an idea of what is being done in Step 4 clearly for three variables second order fuzzy time series. Because of dealing with a second order fuzzy time series, four inputs are employed in neural network model, so that lagged variables $F(t-1),F(t-2),G_1(t-1),\ G_1(t-2),G_2(t-1),G_2(t-2)$ are obtained from three fuzzy time series $F(t),G_1(t)$ and $G_2(t)$. These series are given in Table **1**. The index numbers ($j$) of $A_j$ of $F(t-1)$ and $F(t-2)$ series are taken as input values whose titles are Input-1 and Input-2, The index numbers ($j$) of $B_{i,j}$ of $G_i(t-1),G_i(t-2),i=1,2$ series are taken as input values whose titles are Input-3, Input-4, Input-5, Input-6 in Table **1** for the neural network model. Also, the index numbers of $A_j$ of $F(t)$ series are taken as target values whose title is Target in Table **1** for the neural network model. When the third observation is taken as an example, inputs values for the learning sample $[A_5, A_6, B_{15}, B_{12}, B_{23}, B_{21}]$ are 5, 6, 5, 2, 3 and 1. Then, target value for this learning sample is 2.

**Table 1:** Symbolic representation of a second order fuzzy time series with three variables

| No | $F(t)$ | $G_1(t)$ | $G_2(t)$ | Input 1 $F(t\text{-}1)$ | Input 2 $F(t\text{-}2)$ | Input 3 $G_1(t\text{-}1)$ | Input 4 $G_1(t\text{-}2)$ | Input 5 $G_2(t\text{-}1)$ | Input 6 $G_2(t\text{-}2)$ | Target $F(t)$ |
|----|--------|----------|----------|------|------|------|------|------|------|------|
| 1 | $A_6$ | $B_{12}$ | $B_{21}$ | --- | --- | --- | --- | --- | --- | --- |
| 2 | $A_5$ | $B_{15}$ | $B_{23}$ | --- | --- | --- | --- | --- | --- | --- |
| 3 | $A_2$ | $B_{13}$ | $B_{21}$ | 5 | 6 | 5 | 2 | 3 | 1 | 2 |
| 4 | $A_4$ | $B_{13}$ | $B_{22}$ | 2 | 5 | 3 | 5 | 1 | 3 | 4 |
| 5 | $A_8$ | $B_{14}$ | $B_{28}$ | 4 | 2 | 3 | 3 | 2 | 1 | 8 |
| 6 | $A_7$ | $B_{16}$ | $B_{25}$ | 8 | 4 | 4 | 3 | 8 | 2 | 7 |

**Step 5.** Forecast. Prepare data for forecasting: $F(t+l-1),\ldots,F(t+l-n+1),F(t+l-n)$ and $G_1(t+l-1),\ldots,G_1(t+l-n+1),G_1(t+l-n),\ldots,G_{k-1}(t+l-1),\ldots,G_{k-1}(t+l-n+1),G_{k-1}(t+l-n)$ are taken as the inputs for the trained feed forward neural network and the output from the model is the fuzzy forecast for $F(t+l)$. For example, while the input values for the learning sample are $[A_5, A_6, B_{15}, B_{12}, B_{23}, B_{21}]$ if the output of ANN is 4 then the fuzzy forecast will be $A_4$.

**Step 6.** Defuzzify each fuzzy forecast $F(t+l)$. In this step, the formula proposed by Jilani and Burney [2] is employed. When the output of used feed forward neural network is $A_j$, the defuzzyfied value of this forecast ($t_j$) is calculated as follows:

$$
t_j = \begin{cases}
\dfrac{1+0.5}{\dfrac{1}{a_1}+\dfrac{1}{a_2}} & if \quad j=1 \\[2em]
\dfrac{0.5+1+0.5}{\dfrac{1}{a_{j-1}}+\dfrac{1}{a_j}+\dfrac{1}{a_{j+1}}} & if \quad 2 \le j \le n-2 \\[2em]
\dfrac{0.5+1}{\dfrac{1}{a_{n-1}}+\dfrac{1}{a_n}} & if \quad j=n
\end{cases}
$$

where $a_j$ is the midpoint of the interval with the biggest membership value in $A_j$ which is obtained from the used feed forward neural network in previous step. For example, if the obtained forecast from the used feed forward neural networks is $A_4$, $a_4$ will be equal to the midpoint of the interval $u_4$. In this case, $a_{j-1}$ and $a_{j+1}$ will be equal to the midpoint of the interval $u_3$ and $u_5$, respectively. Jilani and Burney [2] can be seen for details about the formula.

## 5. APPLICATION

We have applied this new technique to the time series data of the total number of annual car road accidents casualties in Belgium from 1974 to 2004 that is given in Table **2**.

**Table 2:** Yearly Car Accident Causalities in Belgium from 1974 to 2004

| Year | Killed | Mortally Wounded | Died within 30 days | Severely wounded | Light casualties |
|------|--------|------------------|---------------------|------------------|------------------|
| 1974 | 1574 | 819 | 2393 | 16506 | 44640 |
| 1975 | 1460 | 701 | 2161 | 15794 | 42423 |
| 1976 | 1536 | 728 | 2264 | 16057 | 44227 |
| 1977 | 1597 | 701 | 2298 | 15830 | 44995 |
| 1978 | 1644 | 728 | 2372 | 16645 | 44797 |
| 1979 | 1572 | 544 | 2116 | 15750 | 42346 |
| 1980 | 1616 | 557 | 2173 | 15915 | 42670 |
| 1981 | 1564 | 454 | 2018 | 15091 | 41915 |
| 1982 | 1464 | 406 | 1870 | 14601 | 40936 |
| 1983 | 1479 | 412 | 1891 | 14864 | 42023 |
| 1984 | 1369 | 363 | 1732 | 14471 | 42456 |
| 1985 | 1308 | 352 | 1660 | 13287 | 39879 |
| 1986 | 1456 | 330 | 1786 | 13764 | 42965 |
| 1987 | 1390 | 380 | 1770 | 13809 | 44090 |
| 1988 | 1432 | 339 | 1771 | 14029 | 44956 |
| 1989 | 1488 | 312 | 1800 | 14515 | 46667 |
| 1990 | 1574 | 190 | 1764 | 13864 | 46818 |
| 1991 | 1471 | 209 | 1680 | 12965 | 43578 |
| 1992 | 1380 | 173 | 1553 | 12113 | 41772 |
| 1993 | 1346 | 171 | 1517 | 11680 | 41736 |
| 1994 | 1415 | 149 | 1564 | 11160 | 40294 |
| 1995 | 1228 | 109 | 1337 | 10267 | 39140 |
| 1996 | 1122 | 115 | 1237 | 9123 | 38390 |
| 1997 | 1150 | 105 | 1255 | 9229 | 39594 |

*Table 2: cont....*

| 1998 | 1224 | 121 | 1345 | 8784 | 41038 |
|------|------|-----|------|------|-------|
| 1999 | 1173 | 126 | 1299 | 8461 | 41841 |
| 2000 | 1253 | 103 | 1356 | 7990 | 39719 |
| 2001 | 1288 | 90  | 1378 | 7319 | 38747 |
| 2002 | 1145 | 118 | 1263 | 6834 | 39522 |
| 2003 | 1035 | 101 | 1136 | 6898 | 42445 |
| 2004 | 953  | 141 | 1094 | 5949 | 41627 |

In this application, the main factor fuzzy time series is the number of killed persons, secondary factor time series are mortally wounded, died within 30 days, severely wounded, light casualties respectively. The application is given step by step below:

***Step 1.*** Universe of discourses and subintervals for five time series are defined as:

$$D_{min} = 953, D_{max} = 1644, \ D_1 = 103, D_2 = 6 \ , \ U = [850,1650] \qquad \textbf{for Killed,}$$

$$D_{min} = 90 \ , D_{max} = 819 \ , D_1 = 40 \ , D_2 = 31 \ , \ V_1 = [50,850] \qquad \textbf{for Mortally Wounded,}$$

$$D_{min} = 1094 \ , D_{max} = 2393 \ , D_1 = 44 \ , D_2 = 57 \ , \ V_2 = [1050,2450] \qquad \textbf{for Died within 30 days,}$$

$$D_{min} = 5949 \ , D_{max} = 16645 \ , D_1 = 149 \ , \ D_2 = 255 \ , \ V_3 = [5800,16900] \qquad \textbf{for Severely wounded,}$$

$$D_{min} = 38390 \ , \ D_{max} = 46818 \ , \ D_1 = 40 \ , D_2 = 282 \ , \ V_4 = [38350,47100] \qquad \textbf{for Light casualties.}$$

The intervals for five time series are defined as given below.

$$\begin{bmatrix} u_1 \\ u_2 \\ u_3 \\ u_4 \\ u_5 \\ u_6 \\ u_7 \\ u_8 \end{bmatrix} = \begin{bmatrix} 850 & 950 \\ 950 & 1050 \\ 1050 & 1150 \\ 1150 & 1250 \\ 1250 & 1350 \\ 1350 & 1450 \\ 1450 & 1550 \\ 1550 & 1650 \end{bmatrix}, \begin{bmatrix} v_{1,1} \\ v_{1,2} \\ v_{1,3} \\ v_{1,4} \\ v_{1,5} \\ v_{1,6} \\ v_{1,7} \\ v_{1,8} \end{bmatrix} = \begin{bmatrix} 50 & 150 \\ 150 & 250 \\ 250 & 350 \\ 350 & 450 \\ 450 & 550 \\ 550 & 650 \\ 650 & 750 \\ 750 & 850 \end{bmatrix}, \begin{bmatrix} v_{2,1} \\ v_{2,2} \\ v_{2,3} \\ v_{2,4} \\ v_{2,5} \\ v_{2,6} \\ v_{2,7} \\ v_{2,8} \end{bmatrix} = \begin{bmatrix} 1050 & 1225 \\ 1225 & 1400 \\ 1400 & 1575 \\ 1575 & 1750 \\ 1750 & 1925 \\ 1925 & 2100 \\ 2100 & 2275 \\ 2275 & 2450 \end{bmatrix},$$

$$\begin{bmatrix} v_{3,1} \\ v_{3,2} \\ v_{3,3} \\ v_{3,4} \\ v_{3,5} \\ v_{3,6} \\ v_{3,7} \\ v_{3,8} \end{bmatrix} = \begin{bmatrix} 5800 & 7100 \\ 7100 & 8500 \\ 8500 & 9900 \\ 9900 & 11300 \\ 11300 & 12700 \\ 12700 & 14100 \\ 14100 & 15500 \\ 15500 & 16900 \end{bmatrix}, \begin{bmatrix} v_{4,1} \\ v_{4,2} \\ v_{4,3} \\ v_{4,4} \\ v_{4,5} \\ v_{4,6} \\ v_{4,7} \\ v_{4,8} \end{bmatrix} = \begin{bmatrix} 38350 & 39400 \\ 39400 & 40500 \\ 40500 & 41600 \\ 41600 & 42700 \\ 42700 & 43800 \\ 43800 & 44900 \\ 44900 & 46000 \\ 46000 & 47100 \end{bmatrix}$$

**Step 2.** Based on defined $U, V_1, V_2, V_3, V_4$ universe of discourses and subintervals, $r_1 = r_2 = 8$ and fuzzy sets $A_j, j = 1, 2, .., 8$ and $B_{ij}, i = 1, 2, 3, 4, j = 1, 2, .., 8$, are defined and given below for main factor and secondary factor time series, respectively. The membership values are:

$$a_{jm} = \begin{cases} 1 & , m = j \\ 0.5 & , m = j - 1 \\ 0.5 & , m = j + 1 \\ 0 & , o.w. \end{cases} , j, m = 1, \ldots, 8,$$

$$b_{ijm} = \begin{cases} 1 & , m = j \\ 0.5 & , m = j - 1 \\ 0.5 & , m = j + 1 \\ 0 & , o.w. \end{cases} , j, m = 1, \ldots, 8, \ i = 1, 2, 3, 4.$$

Then,

$$
\begin{aligned}
A_1 &= a_{11}/u_1 + a_{12}/u_2 + \ldots + a_{1,8}/u_8 & B_{i,1} &= b_{i,1,1}/v_{i,1} + b_{i,1,2}/v_{i,2} + \ldots + b_{i,1,8}/v_{i,8} \\
A_2 &= a_{21}/u_1 + a_{12}/u_2 + \ldots + a_{2,8}/u_8 & B_{i,2} &= b_{i,2,1}/v_{i,1} + b_{i,2,2}/v_{i,2} + \ldots + b_{i,2,8}/v_{i,8} \\
&\quad\vdots & &\quad\vdots \\
A_8 &= a_{8,1}/u_1 + a_{8,2}/u_2 + \ldots + a_{8,8}/u_8 & B_{i,8} &= b_{i,8,1}/v_{i,1} + b_{i,8,2}/v_{i,2} + \ldots + b_{i,8,8}/v_{i,8}
\end{aligned}
, \ i = 1, 2, 3, 4
$$

**Step 3.** The observations are fuzzified. The fuzzified observations for the killed time series are demonstrated in Table **3**.

**Table 3:** Fuzzified observations for the killed time series

| Year | Killed | Fuzzified killed | Year | Killed | Fuzzified killed |
|------|--------|------------------|------|--------|------------------|
| 1974 | 1574 | $A_8$ | 1990 | 1574 | $A_8$ |
| 1975 | 1460 | $A_7$ | 1991 | 1471 | $A_7$ |
| 1976 | 1536 | $A_7$ | 1992 | 1380 | $A_6$ |
| 1977 | 1597 | $A_8$ | 1993 | 1346 | $A_5$ |
| 1978 | 1644 | $A_8$ | 1994 | 1415 | $A_6$ |
| 1979 | 1572 | $A_8$ | 1995 | 1228 | $A_4$ |
| 1980 | 1616 | $A_8$ | 1996 | 1122 | $A_3$ |
| 1981 | 1564 | $A_8$ | 1997 | 1150 | $A_4$ |
| 1982 | 1464 | $A_7$ | 1998 | 1224 | $A_4$ |
| 1983 | 1479 | $A_7$ | 1999 | 1173 | $A_4$ |
| 1984 | 1369 | $A_6$ | 2000 | 1253 | $A_5$ |
| 1985 | 1308 | $A_5$ | 2001 | 1288 | $A_5$ |
| 1986 | 1456 | $A_7$ | 2002 | 1145 | $A_3$ |
| 1987 | 1390 | $A_6$ | 2003 | 1035 | $A_2$ |
| 1988 | 1432 | $A_6$ | 2004 | 953 | $A_2$ |
| 1989 | 1488 | $A_7$ | | | |

**Step 4.** Feed forward ANN is used to determine fuzzy relationships. The inputs of the ANN are being changed according to the order of the model since it is changed between 1 through 4 in the application. For example, the inputs of ANN for the second order five variables (one main factor and four secondary factor time series) model are given as follows:

$$F(t-1), F(t-2), G_1(t-1), G_1(t-2), G_2(t-1), G_2(t-2), G_3(t-1), G_3(t-2), G_4(t-1), G_4(t-2)$$

The target value of ANN is $F(t)$. The outputs of ANN will be the predicted values of the fuzzy time series denoted by $\hat{F}(t)$.

The number of neurons in the hidden layer for the ANN has been determined by changing the number of neurons between 1 through 5. The algorithm is applied to each possible case that is 20 in total. In every stage of ANN, the logistic activation function has been used. Also, the backward learning algorithm is employed in the training stage of the network.

***Step 5.*** The outputs of ANN will be the predicted values of the fuzzy time series denoted by $\hat{F}(t)$.

***Step 6.*** The formula proposed by Jilani and Burney [2] is used to calculate the defuzzyfied forecasts.

For each of the 20 different cases, mean squared errors (MSE) and the average forecasted error rate (AFER) are calculated based on the model order and the number of neurons in the hidden layer. The formulas of MSE and AFER are given below. Table **4** shows different MSE and AFER values that are obtained from changing the model order and the number of neurons in the hidden layer of feed forward ANN.

$$MSE = \sum_{j=1}^{n} \frac{(fv_j - av_j)}{n} \tag{5}$$

$$AFER = \sum_{j=1}^{n} \frac{\left|(fv_j - av_j)/av_j\right|}{n} \times 100\% \tag{6}$$

where $fv_j$ and $av_j$ represent the forecasted value of day $j$ and the actual value of day $j$, respectively.

**Table 4:** MSE and AFER values obtained from changing the number of neurons in the hidden layer and the model order

| | Model Order | | | | | | | |
| | 1 | | 2 | | 3 | | 4 | |
| NHLN | MSE | AFER | MSE | AFER | MSE | AFER | MSE | AFER |
|---|---|---|---|---|---|---|---|---|
| 1 | 5615.429 | 4.908 | 3590.600 | 3.752 | 2204.388 | 2.795 | 1483.608 | 2.377 |
| 2 | 3699.723 | 3.743 | 1152.997 | 2.209 | 1138.900 | 2.196 | 1143.669 | 2.204 |
| 3 | 1356.175 | 2.364 | 1152.997 | 2.209 | 1138.900 | 2.196 | 1143.669 | 2.204 |
| 4 | 1954.353 | 2.676 | 71480.562 | 17.378 | 1138.900 | 2.196 | 1143.669 | 2.204 |
| 5 | 1159.360 | 2.219 | 1152.997 | 2.209 | 1138.900 | 2.196 | 1143.669 | 2.204 |

As can be seen in Table **4**, the smallest value for MSE and AFER obtained from the proposed method are 1138.9 and 2.196, respectively. In our proposed method, the best result has been found for the third order model. When the model order is 3, for all cases in which the number of neurons in the hidden layer is 2, 3, 4 and 5, the same MSE and AFER values have been found. Consequently, the forecasted values obtained from our new techniques for the case with 3rd order model and 2 neurons of the hidden layer are listed in Table **5**. This table also presents the forecasts from the method of Jilani and Burney [2] and Lee *et al.* [3].

It has also seen that the smallest AFER value was obtained from our proposed method which is 2.1962. The real values and the forecasts obtained from all the methods in the literature and the forecasts obtained from our proposed method are shown in Fig. **2**. When Fig. **2** is examined, it is easily seen that our proposed method has performed better than existing methods.

**Table 5:** The comparison of the methods proposed by Jilani and Burney (1-2) [2], Lee *et al.* [3] and the method proposed in this chapter for yearly car accident casualties in Belgium from 1974 to 2004

| Years | Actual Killed | Jilani and Burney (2008) Method 1 | Jilani and Burney (2008) Method 2 | Lee *et al.* (2006) | The Proposed Method |
|---|---|---|---|---|---|
| 1974 | 1574 | --- | --- | --- | --- |
| 1975 | 1460 | --- | --- | --- | --- |
| 1976 | 1536 | --- | --- | --- | --- |
| 1977 | 1597 | 1497 | 1497 | 1500 | 1565.2174 |
| 1978 | 1644 | 1497 | 1497 | 1500 | 1565.2174 |
| 1979 | 1572 | 1497 | 1497 | 1500 | 1565.2174 |
| 1980 | 1616 | 1598 | 1497 | 1500 | 1565.2174 |
| 1981 | 1564 | 1598 | 1497 | 1500 | 1565.2174 |
| 1982 | 1464 | 1498 | 1497 | 1500 | 1496.6592 |
| 1983 | 1479 | 1498 | 1497 | 1500 | 1496.6592 |
| 1984 | 1369 | 1398 | 1497 | 1500 | 1396.4194 |
| 1985 | 1308 | 1298 | 1396 | 1400 | 1296.1424 |
| 1986 | 1456 | 1498 | 1296 | 1300 | 1496.6592 |
| 1987 | 1390 | 1398 | 1497 | 1500 | 1396.4194 |
| 1988 | 1432 | 1398 | 1396 | 1400 | 1396.4194 |
| 1989 | 1488 | 1498 | 1396 | 1400 | 1496.6592 |
| 1990 | 1574 | 1598 | 1497 | 1500 | 1565.2174 |
| 1991 | 1471 | 1498 | 1497 | 1500 | 1496.6592 |
| 1992 | 1380 | 1398 | 1497 | 1500 | 1396.4194 |
| 1993 | 1346 | 1298 | 1396 | 1400 | 1296.1424 |
| 1994 | 1415 | 1398 | 1296 | 1300 | 1396.4194 |
| 1995 | 1228 | 1198 | 1396 | 1400 | 1195.8188 |
| 1996 | 1122 | 1098 | 1095 | 1100 | 1095.4357 |
| 1997 | 1150 | 1198 | 1196 | 1200 | 1195.8188 |
| 1998 | 1224 | 1198 | 1196 | 1200 | 1195.8188 |
| 1999 | 1173 | 1198 | 1196 | 1200 | 1195.8188 |
| 2000 | 1253 | 1298 | 1296 | 1300 | 1296.1424 |
| 2001 | 1288 | 1298 | 1296 | 1300 | 1296.1424 |
| 2002 | 1145 | 1098 | 1095 | 1100 | 1095.4357 |
| 2003 | 1035 | 997 | 995 | 1000 | 994.9749 |
| 2004 | 953 | 997 | 995 | 1000 | 994.9749 |
| **AFER** | | 2.6951 | 5.2444 | 5.2483 | 2.1962 |

## 6. CONCLUSION

Some methods have been proposed in the literature [2, 3] in order to analyze multivariate high order fuzzy time series forecasting models. In this approaches, however, too many computations are needed to determine fuzzy logic relations. In order to avoid intense computations and get more accurate results, we propose a new method based on using ANN to determine fuzzy logic relations and the formula defined by

Jilani and Burney [2] to calculate defuzzyfied forecasts. For the aim of a comparative study, our proposed method and other methods proposed by Jilani and Burney [2] and Lee *et al.* [3] are applied to time series that are yearly car accident causalities in Belgium from 1974 to 2004. As a result of the comparison, the proposed method produces better forecasts than ones obtained from the methods presented by Jilani and Burney [2] and Lee *et al.* [3] in terms of AFER and MSE measures. The proposed method provides two advantages that are more accurate forecasts and easy implementation. In this study, we only use the feed forward ANN. Different architecture types of ANN can also be used to reach more accurate forecasts for future studies.

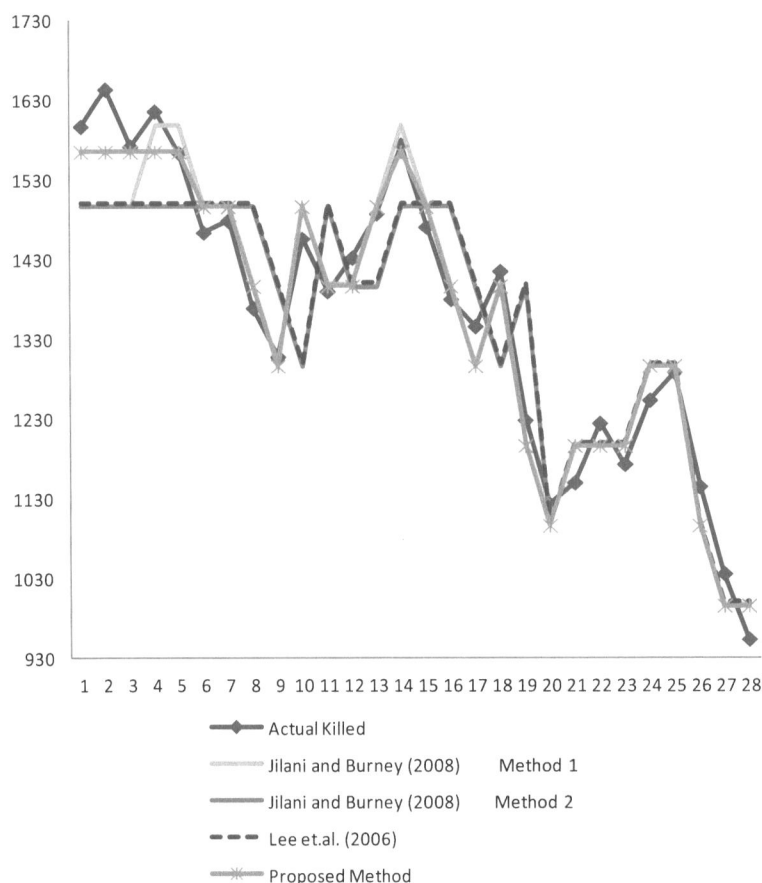

**Figure 2:** The comparison of the proposed method with other methods.

## REFERENCES

[1]   C.H. Aladag, M.A. Basaran, E. Eğrioğlu, U. Yolcu, and V.R. Uslu, "Forecasting in high order fuzzy times series by using neural networks to define fuzzy relations," *Expert Syst Appl*, vol. 36, no. 3, pp. 4228–4231, 2009.

[2]   T.A. Jilani, and S.M.A. Burney, "Multivariate stochastic fuzzy forecasting models," *Expert Syst Appl*, vol. 35, pp. 691-700, 2008.

[3]   L.W. Lee, L.H. Wang, S.M. Chen, and Y.H. Leu, "Handling forecasting problems based on two factors high order fuzzy time series," *IEEE Trans Fuzzy Syst*, 14, 468-477, 2006.

[4]   Q. Song, and B.S. Chissom, "Fuzzy time series and its models," *Fuzzy Set Syst*, vol. 54, pp. 269-277, 1993.

[5]   Q. Song, and B.S. Chissom, "Forecasting enrollments with fuzzy time series - Part I," *Fuzzy Set Syst*, vol. 54, pp. 1-10, 1993.

[6]   Q. Song, and B.S. Chissom, "Forecasting enrollments with fuzzy time series- Part II," *Fuzzy Set Syst*, vol. 62, pp. 1-8, 1994.

[7]   L.A. Zadeh, "Fuzzy Sets," *Inform. and Control*, vol. 8, pp. 338-353, 1965.

[8]    S.M. Chen, "Forecasting enrollments based on high order fuzzy time series," *Cybernet Syst*, vol. 33, pp. 1-16, 2002.

[9]    T.K. Yu, and K. Huarng, "A bivariate fuzzy time series model to forecast the TAIEX," *Expert Syst Appl*, vol. 34, no. 4, pp. 2945-2952, 2008.

[10]   J.M. Zurada, *Introduction of artificial neural systems*, St. Paul: West Publishing, 1992.

[11]   G. Zhang, and B.E. Patuwo, and Y.M. Hu, "Forecasting with artificial neural networks: the state of the art," *Int J Forecasting*, vol. 14, pp. 35-62, 1998.

[12]   E. Eğrioğlu, C.H. Aladag, U. Yolcu, V.R. Uslu, M.A. Basaran, "A new approach based on artificial neural networks for high order multivariate fuzzy time series," *Expert Syst Appl*, vol. 36, no. 7, pp. 10589-10594, 2009.

# Subject Index

**A**

Activation function4, 5, 10-15, 17, 19, 20, 26-30, 33-35, 41, 42, 45, 49, 88, 89, 90, 92, 98, 99, 103, 105, 110, 121, 126.

Akaike Information Criterion 18, 19, 21-24, 35-37, 61.

Artificial neural networks3-8, 11-13, 17, 18-22, 24, 26, 27, 34, 40-46, 48-51, 64, 88-90, 94, 96-99, 105, 108, 109, 111, 112, 116, 118, 119.

Architecture selection 18, 19, 21, 23, 24, 34-37, 51, 88-94.

Autoregressive 11, 18, 26, 56, 88, 96, 97.

Autoregressive fractional integrated moving average 18.

Autoregressive integrated moving average 3, 7, 26, 56-58, 60, 62, 96.

**B**

Back propagation learning method 5, 20, 27, 30, 41, 98, 120, 121.

Bayesian Information Criterion 18, 19, 21-24, 35-37, 61.

Bivariate fuzzy time series forecasting model 6, 96, 97, 118, 119.

Box-Jenkins method 7, 11, 26, 27, 30, 33, 56, 96, 97, 102, 104.

**C**

Combined forecasts 108, 109, 110, 115.

**D**

Difference 14, 20, 22, 27, 49, 61, 64, 68-70, 73, 74, 90, 98, 121.

**E**

Elman's recurrent neural network 7, 11-15, 17, 34.

Entropy 78-83, 85,87,88.

**F**

Feed forward neural networks7, 11, 27, 29, 33, 40, 41, 44, 46, 48-51, 53, 55, 89, 90, 93, 95, 123.

First order fuzzy time series forecasting model 6, 65, 66, 96, 110, 118.

Fuzzy c-means 78,79,82,87.

Fuzzy number 48-51, 55, 57, 59, 60, 62, 83.

**G**

Genetic algorithms 35, 88.

**H**

Heuristic method 90.

Hidden layer 4, 5, 11-18, 22, 23, 27-30, 33-37, 41, 44, 49, 51, 53, 54, 69, 71, 73, 74, 89-91, 93, 98, 103-106, 109, 110, 112, 113, 120, 122, 126.

High order fuzzy time series forecasting model 6, 66, 78, 112, 118, 121, 127.

Hybrid methods 3, 4, 8, 106.

**I**

Interval estimates 56, 60, 62.

**L**

Length of intervals 64, 67-69, 71-73, 75-80, 83, 85-87, 102-106, 111-113.

Levenberg-Marquardt method 12, 20, 27, 28,30, 45, 89, 92, 98, 103, 105.

Long memory 18.

# Author Index